BIOPHYSICS
PRINCIPLES AND TECHNIQUES

BIOPHYSICS
PRINCIPLES AND TECHNIQUES

M A SUBRAMANIAN

Reader
PG and Research Department of Zoology
Chikkaiah Naicker College
Erode - 638 004, TN

MJP PUBLISHERS
Chennai 600 005

MJP Publishers

All rights reserved   No. 44, Nallathambi Street,
Printed and bound in India   Triplicane, Chennai 600 005
MJP 009   © Publishers, 2024

Publisher : C. Janarthanan

PREFACE

The nature of living systems can be explained only in terms of scientific principles. Therefore, in recent times, specialized areas like physics, chemistry, statistics and so on have been incorporated in biology. There is an interrelationship among physical principles, chemical composition and biology of organisms. Hence the branch biophysics has been included in the curriculum of biological sciences. Biophysics deals with the role of physical principles in the organization and functioning of living systems. In this respect, this field can also be considered as a branch of physiology because most of the physiological processes in living organisms is controlled by physical principles.

Owing to the existence of a wide lacuna of textbooks in biophysics, an attempt has been made to bring out this book to fulfill the requirements of students of biology. The book has been devoted to explain various physical principles involved in biological systems along with various techniques. As students of biological sciences may not be well-versed with mathematics, only simple mathematical explanations have been incorporated in the book. The book has been written in simple language without complexity to harmonize the syllabi of various universities and colleges. Standard books, journal materials and review articles have been referred.

The support rendered by my family members especially my wife Jeevarethinam and the research scholars of my department are greatly acknowledged. I am thankful to MJP Publishers, Chennai, for the neat execution and publication of the book. Valuable criticisms and suggestions for future improvement of the book are greatly appreciated.

M A Subramanian

CONTENTS

1. INTRODUCTION 1

   1.1 Introduction 3

2. BIOMOLECULES 5

   2.1 Organisation of Molecules 7
   2.2 Macromolecules and Intermolecular Forces 11
      2.2.1 Stability of Macromolecules 12
      2.2.2 Types of Bonds in Biological Molecules 13
   2.3 Biological Membranes 18
      2.3.1 Lipid Composition 18
      2.3.2 Protein Composition 19
   2.4 Extracellular Matrix 23
      2.4.1 Extracellular Matrix in Epithelial Tissues 24
      2.4.2 Extracellular Matrix in Non-epithelial Tissues 25

PROTEINS 26

   2.5 Organisation of Protein 26
      2.5.1 Primary Structure 27
      2.5.2 Secondary Structure 28
      2.5.3 Tertiary Structure 30
      2.5.4 Quaternary Structure 32
      2.5.5 Cellular Aspects of Protein Folding 32
   2.6 Interaction of Proteins 34
      2.6.1 Protein–Protein Interactions 34
      2.6.2 Protein–Ligand Interactions 35
   2.7 Sequencing of Proteins 38
      2.7.1 Determination of Amino Acid Terminals 38
      2.7.2 Protein Sequencing through Determination of Gene Sequences 38

		2.7.3 Edman Degradation Method	39
		2.7.4 Sequenators	39
	2.8	Protein Denaturation	41
		2.8.1 Bonds in Protein Molecules	41

CARBOHYDRATES — 45

	2.9	Organisation of Carbohydrates	45
		2.9.1 Monosaccharides	45
		2.9.2 Disaccharides	48
		2.9.3 Polysaccharides	49
		2.9.4 Glycoproteins	51

LIPIDS — 53

	2.10	Organisation of Lipids	53
		2.10.1 Phospholipids	55
		2.10.2 Lipoproteins	57

NUCLEIC ACIDS — 58

	2.11	Organisation of Nucleic Acids	58
		2.11.1 Primary Structure of DNA	61
		2.11.2 Secondary Structure of DNA	61
		2.11.3 Tertiary Structure of DNA	65
		2.11.4 Structure of RNA	66
		2.11.5 Sequencing of Nucleic Acids	67
		2.11.6 Nucleoproteins	69
	2.12	Macromolecules of Immune System	70
		2.12.1 Antigens	71
		2.12.2 Antibodies	72
	Review Questions		74

3. PRINCIPLES OF KINETICS OF MOLECULES — 77

	3.1	Diffusion	79
		3.1.1 Factors Affecting Diffusion	79
		3.1.2 Simple Diffusion	80
		3.1.3 Fick's Law of Diffusion	82
		3.1.4 Diffusion of Electrolytes	82
		3.1.5 Facilitated or Accelerated Diffusion	83
		3.1.6 Biological Significance of Diffusion	84

3.2   Osmosis   85
      3.2.1   Osmotic Pressure   86
      3.2.2   Laws of Osmosis   86
      3.2.3   Determination of Osmotic Pressure (Osmometry)   88
      3.2.4   Determination of Concentration by Depression in
              Freezing Point Method   89
      3.2.5   Determination of Osmotic Pressure of Electrolytes   91
      3.2.6   Determination of Molecular Weight by
              Hydrodynamic Method   91
      3.2.7   Biological Significance of Osmosis   92
3.3   Filtration   92
      3.3.1   Passage of Fluid through
              Blood Vessels into Tissue Space   93
      3.3.2   Formation of Urine by Filtration   93
      3.3.3   Role of Filtration in Microbiology   94
3.4   Dialysis   94
      3.4.1   Principle of Dialysis in Artificial Kidney   96
      3.4.2   Kinds of Dialysis   97
3.5   Surface Tension   98
      3.5.1   Kinetic Theory of Surface Tension   98
      3.5.2   Factors Affecting Surface Tension   99
      3.5.3   Determination of Surface Tension of
              Liquids by Capillary Rise Method   100
      3.5.4   Determination of Surface Tension of
              Liquids by Drop Weight Method   101
      3.5.5   Determination of Interfacial
              Surface Tension of Liquids   102
      3.5.6   Biological Significance of Surface Tension   103
3.6   Adsorption   103
      3.6.1   Factors Affecting Adsorption   104
      3.6.2   Adsorption of Ions by Solids   104
      3.6.3   Adsorption of Gases by Solids   105
      3.6.4   Adsorption of Ions by Liquids   105
      3.6.5   Adsorption and Colloids   106
      3.6.6   Biological Significance of Adsorption   108
3.7   Hydrotropy   109
      3.7.1   Biological Importance of Hydrotropy   110

3.8   Precipitation   111
       3.8.1   Biological Significance of Precipitation   112
3.9   Viscosity   112
       3.9.1   Factors Affecting Viscosity   114
       3.9.2   Determination of Viscosity of Liquids   114
       3.9.3   Biological Significance of Viscosity   117
3.10   Colloids   118
       3.10.1   Types of Colloids   118
       3.10.2   Characteristics of Colloids   119
       3.10.3   Stability of Colloids   121
       3.10.4   Gel   121
       3.10.5   Emulsions   121
       3.10.6   Techniques for the Separation of Colloids   122
       3.10.7   Biological Importance of Colloids   122
3.11   Gibb's Donnan Equilibrium (Donnan
       Membrane Equilibrium)   123
       3.11.1   Donnan Equilibrium in Living Systems   125
       Review Questions   *127*

4.   **PRINCIPLES OF OPTICS IN
     BIOLOGICAL STUDIES**   **129**

4.1   Characteristics of Light   131

MICROSCOPY   133

TYPES OF MICROSCOPES   134

4.2   Compound Microscope   134
4.3   Phase Contrast Microscope   136
       4.3.1   Optical Principle   137
4.4   Interference Microscope   138
4.5   Polarization Microscope   140
4.6   Ultraviolet Microscope   141
4.7   Fluorescent Microscope   143
4.8   Ultramicroscope (or) Dark-field Microscope   144
4.9   Electron Microscope   145
       4.9.1   Transmission Electron Microscope (TEM)   146
       4.9.2   Scanning Electron Microscope (SEM)   148

4.9.3   Electron Cryomicroscopy (EC)   149
4.9.4   Scanning Tunneling Electron Microscope (STEM)   149

PHOTOMETRY OR ABSORPTIOMETRY   151

4.10   Important Components in Instruments of Photometry   154
    4.10.1   Light Source   154
    4.10.2   Monochromator   154
    4.10.3   Sample Holders   156
    4.10.4   Light Sensitive Detectors   156
4.11   Colorimeter   157

SPECTROPHOTOMETER   158

4.12   Ultraviolet and Infrared Spectrophotometers   159
4.13   Flame Photometer   160
4.14   Atomic Absorption Flame Photometer   162
Review Questions   *164*

5.   BIOPHYSICAL PHENOMENA IN BIOCHEMICAL STUDIES   **167**

5.1   Hydrogen Ion Concentration (pH)   169
    5.1.1   pH Scale   170
    5.1.2   Determination of pH   173
5.2   pH Meter   175
    5.2.1   Principle   175
    5.2.2   Electrode System in pH Meter   176
    5.2.3   Factors Affecting Measurement of pH   182
5.3   Importance of Buffers in Biological Systems   183
    5.3.1   Bicarbonate Buffer System   184
    5.3.2   Phosphate Buffer System   184
    5.3.3   Protein Buffer System   185
    5.3.4   Amino Acid Buffer System   185
    5.3.5   Buffer System in Human Blood   185
    5.3.6   pH-dependent Ionization of Biomolecules   187
5.4   Centrifugation   189
    5.4.1   Basic Principle of Centrifugation   189

CENTRIFUGES   191

5.5   Desktop Centrifuge or Clinical Centrifuge   192

5.6	High-speed Centrifuge	192
5.7	Ultracentrifuge	192
	PREPARATIVE ULTRACENTRIFUGE	193
5.7.1	Differential Centrifugation	193
5.7.2	Density Gradient Centrifugation	194
	ANALYTICAL ULTRACENTRIFUGE	196
5.7.3	Determination of Molecular Weight by Ultracentrifugation	197
5.8	Chromatography	198
5.8.1	Principles of Chromatography	198
	TYPES OF CHROMATOGRAPHY	200
	ADSORPTION CHROMATOGRAPHY	200
5.9	Paper Chromatography	200
5.9.1	Choice of Paper	201
5.9.2	Application of the Sample on the Paper	201
5.9.3	Selection of Solvents	202
5.9.4	Development of Chromatogram	203
5.9.5	Location of Spots	204
	TWO-DIMENSIONAL PAPER CHROMATOGRAPHY	205
5.10	Thin Layer Chromatography (TLC)	206
5.11	Column Chromatography	207
5.12	Gel-permeation Chromatography	208
5.12.1	Determination of Molecular Weight of Macromolecules by Gel Permeation Chromatography	209
5.13	Ion-exchange Chromatography	210
5.14	Affinity Chromatography	211
5.15	Gas-liquid Chromatography (GLC)	212
5.16	High Performance Liquid Chromatography (HPLC)	213
5.17	Electrophoresis	215
5.17.1	Principle 215	
5.17.2	Factors Affecting the Migration of Substances	216
5.17.3	Supporting Media in Electrophoresis	217

TYPES OF ELECTROPHORESIS   218

5.18   Boundary Electrophoresis   218

ZONE ELECTROPHORESIS   219

5.19   Paper Electrophoresis   220
5.20   Cellulose Acetate Electrophoresis   222
5.21   Gel Electrophoresis   222
5.22   Polyacrylamide Gel Electrophoresis (PAGE)   222
5.23   SDS–Polyacrylamide Electrophoresis (SDS–PAGE)   225
5.24   Slab Gel Electrophoresis   225
5.25   Two-dimensional Gel Electrophoresis   226
5.26   Immunoelectrophoresis   227
Review Questions   *228*

6. ELECTROMAGNETIC RADIATION AND SPECTROSCOPY IN BIOLOGICAL STUDIES   231

6.1   Electromagnetic Radiation   233
    6.1.1   Wave Properties of Electromagnetic Radiation   233
    6.1.2   Particle Properties   234
    6.1.3   Electromagnetic Spectrum   234

SPECTROSCOPY   235

6.2   Ultraviolet and Visible Spectroscopy   237
6.3   Infrared Spectroscopy   239
6.4   Fluorescent Spectroscopy   242

ATOMIC SPECTROSCOPY   243

6.5   Atomic Absorption Spectroscopy   243
6.6   Double Beam Atomic Absorption Spectroscopy   245
6.7   Atomic Emission Spectroscopy   245
6.8   Mass Spectroscopy   246
6.9   Raman Spectroscopy   249

X-RAY SPECTROSCOPY   251

6.10   X-ray absorption spectroscopy   252
6.11   X-ray Diffraction Spectroscopy (X-ray Diffraction Crystallography)   253

6.11.1	Transmission Photographic Method	255
6.11.2	X-ray Ionization Spectrometer	256
6.11.3	Rotating Crystal Diffraction Spectrometer	256
6.11.4	Powder Crystal Diffraction Spectrometer	257
6.12	Nuclear Magnetic Resonance (NMR) Spectroscopy	258
6.13	Electron Spin Resonance (ESR) Spectroscopy	260
	Review Questions	262

7. OTHER OPTICAL TECHNIQUES IN BIOLOGICAL STUDIES   265

7.1	Optical Rotatory Dispersion (ORD) and Circular Dichroism (CD)	267
7.1.1	Optical Activity	267
7.1.2	Optical Rotatory Dispersion (ORD)	268
7.2	Circular Dichroism	269
7.2.1	Measurements	269
7.2.2	Measurement of ORD	270
7.2.3	Measurement of CD	270
7.3.	LASER (Light Amplification by Stimulated Emission of Radiation)	272
7.3.1	Principle	272
7.3.2	Instrumentation	273
7.3.3	The Ruby Laser	273
7.3.4	Helium–Neon laser	274
	Review Questions	275

8. BIOELECTRICITY AND NERVE IMPULSE CONDUCTION   277

8.1	Membrane Potential	279
8.2	Resting Membrane Potential	280
8.3	Action Potential and Nerve Impulse Conduction	281
8.4	Rate of Nerve Impulse Conduction	285
8.5	Recording of Nerve Impulses	286
8.5.1	Resting Membrane Potential	288
8.5.2	Injury Potential	289

8.5.3 Monophasic Action Potential 289

8.5.4 Diphasic Action Potential 290

Review Questions 291

9. RADIATION BIOLOGY 293

9.1 Radioactivity 295

9.1.1 Natural Radioactivity 295

9.1.2 Artificial or Induced Radioactivity 296

9.1.3 Radioactive Disintegration 296

9.1.4 Units of Radioactivity 296

MEASUREMENT OF RADIOACTIVITY 297

9.2 Geiger-Muller Counter (GM tube) 297

9.3 Proportional Counter 298

9.4 Scintillation Counter 299

9.5 Liquid Scintillation Counter 299

9.6 Crystal Counter 300

9.6.1 Method of Detection of Disintegration Frequency 300

9.7 Autoradiography 301

9.7.1 Specimen Preparation for Autoradiography 302

9.7.2 Methods to Make Contact Between Specimen
and Photographic Emulsion 302

9.8 Biological Effects of Radiation 304

Review Questions 304

Glossary 307

References 315

Index 319

INTRODUCTION

1.1 INTRODUCTION

Biophysics is an interdisciplinary science which applies the principles of physics to the subject of biology in order to increase the understanding of biological systems. In other words, the methods invented for physical challenges are integrated in biological sciences to solve the biological problems. In general, atoms are aggregated to form molecules through interaction of electrons. A variety of interactions in the form of interparticle forces occur between the atoms in order to stabilize the biological molecules. These molecules undergo polymerization resulting in the formation of macromolecules with various conformational changes. Proteins, polysaccharides and nucleic acids are the major biological polymers which are necessary for the existence of life. The biological membranes are nothing but lipoproteins, the permeability of which is determined by various physical factors.

In living organisms, the cells are grouped to form tissues which in turn are organized into organs and organ systems. These cells perform different functions and are so specialized for division of labour. At present, major mysteries of various physiological functions in living organisms are being unravelled in terms of physical concepts, as most of the physiological processes in living systems are governed by the principles of physics. Moreover, molecular biology involves the identification and separation of biological macromolecules by using several techniques which obey the laws of physical and chemical phenomena. The functions of biological macromolecules depend on their structures, thus it is of immense importance to know their three-dimensional structures, so as to understand the mechanism of their functions. In this way, the study of biomolecules necessitates several physical and chemical techniques.

As biological structures are too small to be visible to the naked eye, microscopes are the versatile tools for biologists.

Chromatography and electrophoresis are separation techniques for qualitative and quantitative determinations of biological substances. Centrifugation is meant for the separation of organic substances and intracellular organelles as well as for the determination of molecular weight of macromolecules. The basis of spectroscopy involves the interaction of electromagnetic radiation with matter. Spectroscopy and NMR application provide information regarding the stereochemistry and conformation of biological molecules. X-ray crystallography is the technique which is highly useful to obtain a detailed and accurate structure of macromolecules. Three-dimensional images of macromolecules are obtained by Laser. The radioactivity emitted by radioisotopes is used in autoradiography, which is employed in the study of cell metabolism.

On the ground of the above facts, it is conceivable that biophysics has been extended as the major arena in the field of biology to understand life processes at molecular level.

BIOMOLECULES

2.1 ORGANISATION OF MOLECULES

An atom consists of a nucleus surrounded by negatively charged electrons, which revolve in definite orbits. The nucleus contains positively charged protons and neutral neutrons. In general, an atom possesses a neutral charge because the negative charges of electrons counterbalance the positive charges of protons. The atom is the smallest part of an element, which cannot be split into simpler substances. The stability of elements is due to their outer electrons, i.e., the elements acquire the electronic configuration of the nearest inert element to become stable. For example, sodium, whose atom possesses a single electron in the outermost orbit, readily gives the solitary electron to have a closed shell distribution and becomes positively charged sodium ion. A molecule is the smallest part of an element in which the atoms combine through chemical bonds. The living systems are formed by a limited number of atoms which combine to form molecules ranging from simple molecules to macromolecules.

The structure and shape of the molecules are determined by the interparticle forces in the form of chemical bonds between the atoms. A molecule with two equal positive and negative charges separated by a small distance will form a dipole as shown in the Figure 2.1. Thus the spatial separation of two equal electrical charges of opposite sign in a molecule is called a **dipole**. An electric dipole is thus formed when the centres of positive and negative charges are separated. In this way a dipole has a positive pole and a negative pole, and this causes an electrical asymmetry in the molecule. Occurrence of an electrical asymmetry in a dipole is called **polarity**. Presence of equal positive and negative charges in a polar molecule separated by a unit distance is called **dipole moment**. Therefore, the potential developed by the electrical symmetry is the dipole moment which depends on the distance between the two opposite charges (l) and the number of charges at each pole (e).

$$\text{Dipole moment } (\mu) = e \times l$$

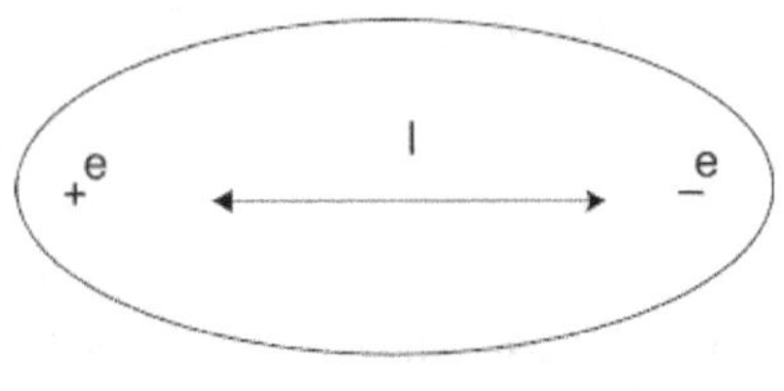

(l - Distance between poles)

Figure 2.1. A dipole.

The dipole moment represents the degree of polarity of a molecule, which is expressed in electrostatic units (esu) and the distance in Angstrom.

In a system where the molecules are in free movement, the positive end of a molecule will be attracted towards the negative end of a neighbouring molecule or vice versa. On the other hand, in an electrically charged medium, the molecules arrange themselves in such a way that the positive ends are directed towards the negative pole and the negative ends are directed towards the positive pole. The bond formed between the elements, which have equal electronegative poles, is called **nonpolar** whereas the bond between the elements, which have different electronegative poles, is called **polar**. In an electric field, the electrons of a nonpolar molecule as well as the positive nuclei are displaced towards the positive pole causing **induced dipoles**. The dipole moments of these molecules are independent of the influence of temperature. On the other hand, the polar molecules are **permanent dipoles,** as they possess permanent positive and negative ends. The dipole moments of these molecules depend on the influence of temperature. Symmetrical molecules do not show dipole moment as the individual bond moments cancel each other. Similarly, the linear molecules also exhibit no dipole moments as the dipole moment on one side cancels the dipole moment on other side of the molecule.

The length of the molecule influences the separation of whole electron charges so that molecules with large dipole moments are highly polar. In aqueous medium, a permanent dipole can be attracted by a nearby ion (**charge–dipole interaction**) or by another permanent dipole. This is because the dipole interactions depend on the orientation of the dipoles. The molecules without dipole moments can be converted into dipoles by electricity. These induced dipole molecules are called as **polarizable molecules**. The interactions of the polarizable molecules are called **dipole–induced interactions**. These interactions involve an anion or a cation in a polarizable molecule so that they are attracted to it (**charge-induced dipole interaction** or **dipole–dipole interaction**). Even two molecules without a net charge or a permanent dipole moment can attract each other if they are brought very close to each other. In these molecules, the distribution of charge is not static but fluctuates. The charge

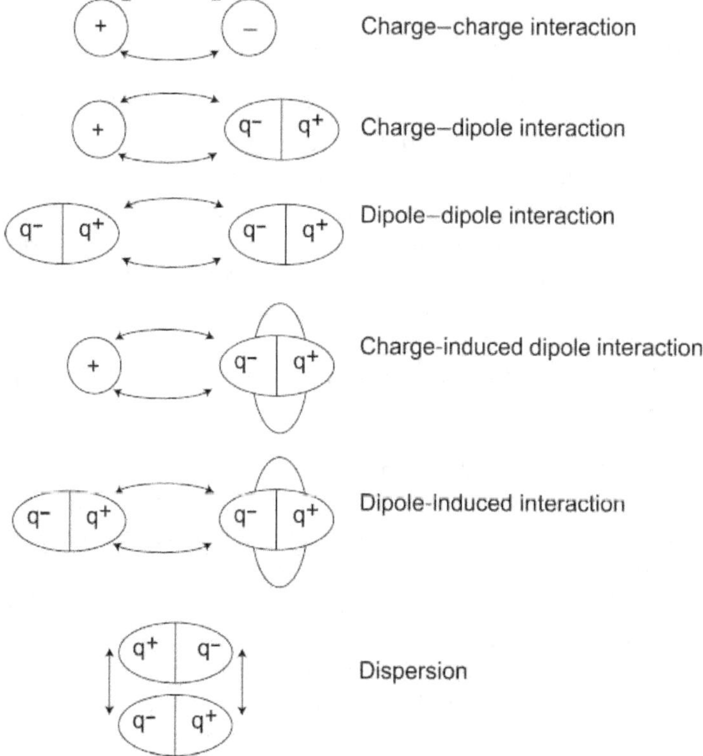

Figure 2.2. Dipole interactions.

fluctuation in these molecules is synchronized with a net attractive force. Such intermolecular forces are called **dispersion forces**. These dipole interactions are shown in Figure 2.2.

When two unidentical atoms are combined together by a covalent bond, the electrons move towards the atom, which has more electronegative charge. As a result, an electrical polarity is developed due to noncoincidence of positive and negative centres. If these centres coincide with each other, then the molecule becomes neutral. Thus the molecules which possess electrical polarity due to positive and negative charges separated by a small distance are called **polar molecules,** and these are dipoles exhibiting dipole moment (OH, NH_2, COOH, etc.) These molecules are otherwise called as **hydrophilic groups** as they have orientation towards water molecules. Electrolytes and amino acids like aspartic acid, glutamic acid and lysine are ionized polar molecules in which the electrostatic attraction between anions and cations are reduced and are surrounded by water molecules. Glucose, glycerol and soluble amino acids like threonine, cysteine and serine are unionized polar molecules which readily dissolve in water and other polar solvents such that they associate with the polar groups of solvent molecules through hydrogen bonds. In short, the molecules with large dipole moments are capable of hydrogen bonding with water or capable of interaction with water dipoles or hydrophilic substances as they readily dissolve in water.

Saturated hydrocarbons, triglycerides, steroids, etc. are **nonpolar molecules** as they are neutral molecules with electrical symmetry where the positive and negative centres coincide without having dipole moment. These molecules are otherwise called as **hydrophobic groups** because they are insoluble in water but soluble in nonpolar solvents like carbon tetrachloride, benzene, ethylene, etc. The nonpolar molecules exhibit no polarity, dipole moment and hydrogen bonding with water due to the presence of electrical symmetry. However the hydrophobic substances are capable of forming aggregates in water.

The molecules, which possess both hydrophilic polar groups and hydrophobic nonpolar groups are called **amphipathic molecules**. As these molecules have polar groups, a small portion remains as dispersed particles in solution. The hydrophilic polar groups of these molecules orient towards water and combine with water molecules whereas the hydrophobic groups orient away from water. In general an amphipathic molecule has a strong hydrophilic head group combined to a hydrophobic tail group. These molecules play an important role in the formation of biological membranes.

The polar molecules orient themselves under electric field due to their dipole moment in which the positive ends direct towards the negative side and the negative ends towards the positive side. In the field of attraction, the polar molecules orient themselves in such a way that the positive pole of one molecule comes nearer to the negative pole of another molecule. In permanent dipole molecules, the force of attraction is inversely proportional to the distance between them. However, the dipole molecules do not move freely as the dipole forces are too small. Moreover, the orientation of the dipole molecules depends on temperature, intermolecular forces and molecular weight and shape of the molecules.

The dipole nature of molecules plays an important role in biological systems. The occurrence of polar side chains in biological molecules makes them soluble in water. Nonpolar lipid molecules are insoluble in water but soluble in nonpolar solvents. The hydrophobic nature of fats renders them to be stored in fat depots and body tissues. Amphiphilic biomolecules are responsible for the formation of lipid bilayers in biological structures. The structure and shape of the molecules and intermolecular forces determine the viscosity of the substances.

2.2 MACROMOLECULES AND INTERMOLECULAR FORCES

Macromolecules are giant molecules formed by many repeat units and are otherwise called as **polymers**. The repeat units are produced by the reaction between two or more monomers leading to **polymerization** in the formation of a macromolecule. The macromolecules may either be natural or synthetic. A polymer chain consists of a large number of repeat units, which are chemically linked together in a variety of ways. The coiling of a flexible chain is regular so as to form a helix or a double helix as in many biopolymers. In biological systems, limited number of atoms combine to form molecules ranging from simple molecules to macromolecules. The small molecules are the building blocks from which the macromolecular aggregates are formed. These molecules are formed by interactions between the atoms by covalent bonds and non-covalent bonds which bring stabilization of groups of atoms within and between molecules.

2.2.1 Stability of Macromolecules

The elements achieve an electronic configuration by losing, gaining or sharing electrons thereby entering into electrovalent, covalent or coordinate covalent linkages respectively. The energy released at the time of overlapping of atomic orbitals to form bonds is called **bond energy** or **stabilized ion energy** which depends on the extent of overlapping of atomic orbitals, the size of the nuclei of atom and the type of bond formed. For example, if the overlapping is more, then more energy is released and the strength of the bond is higher. The distance between two nuclei is reduced by a greater overlap so that the interatomic distance is shortened. But the rate of overlapping decreases when the positively charged nuclei of two atoms repel each other. Now any more closeness increases the repulsion of the atoms so that the molecule becomes unstable. Therefore, the atoms cannot come together beyond a specific distance where the forces of repulsion become exactly equal to the forces of attraction for overlapping. This equilibrium distance between the two nuclei is called **bond length** or **bond distance**.

The function of biological molecules are determined by their structure which in turn is determined by the forces acting between the atoms of a molecule. The forces between the atoms of molecule may be strong or weak; strong forces are responsible for the primary structure of biological molecules whereas weak forces decide the secondary, tertiary and quaternary structures. Thus the chemical or molecular structures of the biological molecules are due to the strong forces whereas the three-dimensional structure is due to the weak forces. If two bonded atoms exhibit variation to attract electrons, they constitute a dipole possessing a dipole moment. In general, the molecular structure of a biological substance is due to the three-dimensional arrangement of its molecules. The configuration of molecules refers to the arrangement of various groups of atoms attached to a centre. Therefore, a change in the configuration of the molecule involves the breakage of bonds and their reformation. The conformation of molecules refers to the arrangement of atoms due to the rotation of part of the molecules. A change in the conformation of the molecule involves rotation and orientation of atoms.

2.2.2 Types of Bonds in Biological Molecules

In biological systems, smaller molecules participate in metabolism whereas the larger molecules are structural or informational in nature. The major biological molecules include water and organic substances such as proteins, polysaccharides and lipids as well as nucleic acids which are in turn made up of simple organic elements namely carbon, oxygen, nitrogen and hydrogen. Proteins, polysaccharides and nucleic acids which are polymers formed by monomers namely amino acids, monosaccharides and nucleotides respectively are the most important. The biological molecules are held together by strong and weak forces between their atoms facilitating interactions in them. The **covalent bond** is a strong interaction in which two atoms share one or multiple pairs of electrons. The **non-covalent bonds** are weak attractive forces but they determine the properties and functions of biomolecules. The non-covalent interactions include four types namely ionic interactions, hydrogen bonds, van der Waal's interactions and hydrophobic effects.

Covalent bonds   In biomolecules, the atoms such as hydrogen, oxygen, carbon, nitrogen, phosphorous and sulphur are most abundant and they readily form covalent bonds with other atoms. These bonds are very stable since a great amount of energy is required to break covalent bonds. A carbon atom usually forms four covalent bonds with other atoms in biological building blocks. A carbon atom which forms linkage with four dissimilar atoms in a non-planar configuration is called asymmetric carbon atom, which can be arranged three-dimensionally in two different ways. As a result, molecules which are mirror images of each other are produced. This property of molecules is called **chirality** and the molecules are called **optical isomers** or **stereoisomers**. Most of the biomolecules have at least one asymmetric carbon atom. In a bond between a carbon atom and three other atoms, all the atoms are in a common plane in which the carbon atom forms two single bonds with two atoms and a double bond with the third atom. The single bonded atom can rotate freely on its axis whereas the double bond cannot rotate. Table 2.1 gives the number of covalent bonds formed by the common atoms.

Table 2.1. Number of covalent bonds by the common atoms.

Atoms	No. of covalent bonds
Hydrogen	1
Oxygen	2
Sulphur–Hydrogen Sulphide	2
–Sulphuric acid	6
Nitrogen–Ammonia	3
–Ammonium ion	4
Phosphorus	5

In a covalent bond, since the bonded atoms exhibit differential attraction towards the electrons, there is an unequal sharing of electrons. The ability of atoms to attract electrons is called its electronegativity. A nonpolar bond is one in which the atoms have similar electronegativities and the electrons are equally shared between two atoms. On the other hand, a polar bond is one in which two atoms possess different electronegativities. For example in O–H bond of water molecule the oxygen atom has greater electronegativity than the hydrogen atom and so the electrons spend more time around the oxygen atom than around the hydrogen atom. Therefore, the O–H bond is said to possess an **electric dipole**.

Non-covalent bonds   The non-covalent bonds are weak bonds since only little energy is required to break their bonds. However the occurrence of a large number of non-covalent bonds can produce highly stable molecules or macromolecules.

Ionic interactions or ionic bonds or electrostatic bonds   The ionic bonds are due to the attraction between oppositely charged ions, that is attraction of a positively charged ion (cation) by a negatively charged ion (anion) or vice versa. The electrostatic bonds do not show any fixed or specific geometric orientation because the electrostatic force around the ions is uniform in all directions. In any biological solution, the important ions are sodium, potassium, calcium, magnesium and chlorides which do not exist in free state but are surrounded by water molecules through ionic interactions as shown in the Figure 2.3. Most of the ionic compounds of biological systems are readily soluble in water due to greater energy by binding of ions with the water dipole. The interaction of two ions in an aqueous medium depends on the concentration of other ions in solution. A higher concentration of other ions causes more ionic interactions in solutions.

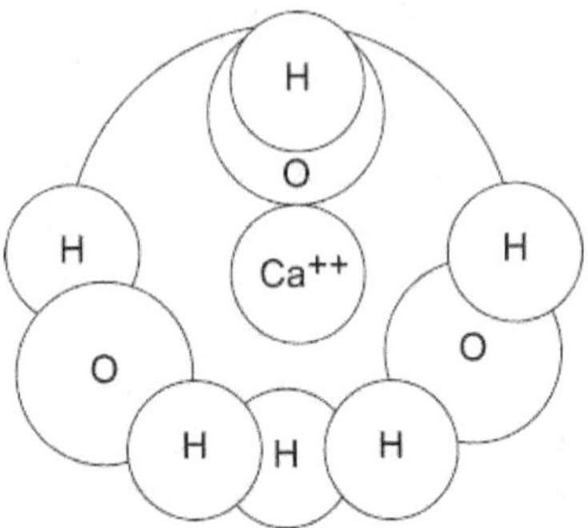

Figure 2.3. Ionic bond between calcium and water.

Hydrogen bonds   Hydrogen bonds are predominant in living systems as the living cells contain 70–90% water. Almost all the biological molecules occur in aqueous environment and so the role of water molecules is very important in the structure, conformation and arrangement of biopolymers. The intermolecular hydrogen bonds are formed within the biopolymers as well as with the adjoining water molecules. Almost all the polar molecules are linked through a network of hydrogen-bonded water molecules in living systems.

The attractive force which causes the binding of hydrogen atom of one molecule with an electronegative atom of another molecule, is called **hydrogen bond** or **proton bond.** The hydrogen bonds are longer and weaker than the covalent bonds but possess directionality. When a hydrogen atom is covalently combined with a strong electronegative atom like oxygen, the positively charged hydrogen atom is attracted towards the negative end of oxygen, thus weakly binding the two molecules. Thus, a number of molecules combine together to form large groups of molecules where the hydrogen acts as a bridge between the electronegative atoms. In a hydrogen bond, both the hydrogen atom and the acceptor atom lie in a straight line. But the non-linear hydrogen atoms are weaker than the linear hydrogen atoms and are capable of stabilizing the macromolecules when they occur in large numbers.

Generally, the hydrophilic molecules with polar ends easily form hydrogen bonds with water and so they are readily soluble in water.

Hydrogen bonding may be either intermolecular or intramolecular. The **intermolecular hydrogen bonding** causes the association of molecules and is influenced by the shape of the molecule. This type of bonding brings changes in number, mass, shape and electronic structure of molecules. The molecular weight of the molecules is increased by the intermolecular hydrogen bonds causing higher melting and boiling points as well as increased viscosity and greater dipole moments. The **intramolecular hydrogen bonding** causes no molecular association but modifies the electronic structure of molecules such that the physical properties of the system remain unaltered. Moreover, this type of bonding reduces the solubility of molecules in water. The occurrence of a large number of intermolecular hydrogen bonding causes considerable folding in macromolecules.

van der Waal's interactions   When two atoms come in close contact with each other, the electrons of one atom will alter the electrons of another atom so that a transient dipole is produced in the second atom. Now the two atoms link together by a non-specific force called **van der Waal's interaction.** This force produces random fluctuations in the distribution of electrons in an atom causing a transient unequal distribution of electrons. These interactions occur in both polar and nonpolar molecules but the strength of the interactions decreases with the increasing distance between atoms.

Hydrophobic effects   The nonpolar molecules are hydrophobic in nature and so are insoluble in water. But they aggregate in water due to hydrophobic interaction in which the water molecules form rigid pentagons or hexagons around the nonpolar molecules. This is because the water molecules cannot form hydrogen bonds with nonpolar molecules. When the nonpolar molecules aggregate in a solution with their hydrophobic surfaces facing each other, the surface area of the nonpolar molecules is reduced. Now the formation of aggregates requires only less quantity of water. In biological systems, the hydrocarbons are insoluble in water because the nonpolar bonds are the covalent bonds between two carbon atoms or between carbon and hydrogen atoms.

2.3 BIOLOGICAL MEMBRANES

All living cells are enveloped by a plasma membrane. In prokaryotes, there is no internal membrane whereas the cell is partitioned to form smaller subcompartments called **organelles** in eukaryotes. Membranes also surround the organelles. The cytoplasm of the cell is the outside part of the organelle and the cytosol forms the aqueous part of the cytoplasm. The biomembranes control the movement of substances from inside to outside of the cell and of the organelles. Invariably, all biomembranes are formed by lipoprotein complex and are amphipathic in nature involving in the regulation of flow of materials as well as in forming the sites for a number of intracellular metabolic reactions. According to fluid-mosaic model given by Singer and Nicalson, the biological membrane is a fluid bilayer formed by complex amphipathic lipids with proteins embedded in or traversing the bilayer. Thermodynamically, the biomembranes are stable as the bilayers are involved in both hydrophobic and hydrophilic interactions in an aqueous medium.

2.3.1 Lipid Composition

The biomembranes are found to contain phospholipids occurring in the form of bilayer with hydrophilic heads and hydrophobic cores or tails as shown in the Figure 2.4. This bilayer is highly stable due to hydrophobic and van der Waal's interactions and is impermeable to hydrophilic substances and ions. In general, the major lipids of biological membranes include phosphoglycerides, sphingolipids and steroids, which are all amphipathic molecules with a polar head group (hydrophilic group) and a nonpolar tail group (hydrophobic). The lipid composition of biological membranes varies greatly. For example, the phospholipids and sphingolipids are asymmetrically distributed in the two layers whereas the cholesterol is evenly distributed. While sphingolipids and cholesterol decrease the membrane fluidity, phosphoglycerides increase the fluidity of the membrane.

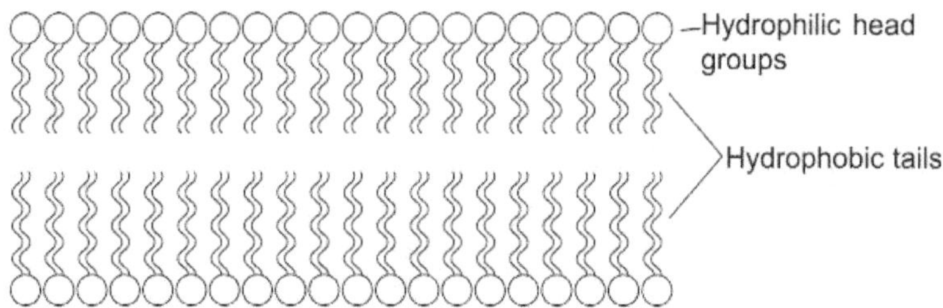

Figure 2.4. Phospholipid bilayer of the biological membrane.

The lipid molecules in a bilayer can rotate freely around their long axis and can diffuse laterally within two layers. As the movements of these molecules are either rotational or lateral, the fatty chains remain in the hydrophobic interior of the membrane. This indicates that the lipids act as fluid in biological membrane in which the fluidity depends on the composition of lipids, hydrophobic tails and temperature. The hydrophobic effect and van der Waal's interactions make the nonpolar tails of phospholipids to aggregate.

2.3.2 Protein Composition

In biological membranes, proteins are associated with the surface of phospholipid bilayer, which provides a hydrophobic condition for the membrane proteins. Generally, the membrane proteins are asymmetrically oriented with respect to the membrane faces. Some proteins occur within the lipid bilayer and others are associated with the cytosolic region of the bilayer. Based on the nature of the membrane- protein interact ions, the membrane proteins are of three types namely integral membrane proteins or transmembrane proteins, lipid-anchored membrane proteins and peripheral membrane proteins as shown in Figure 2.5.

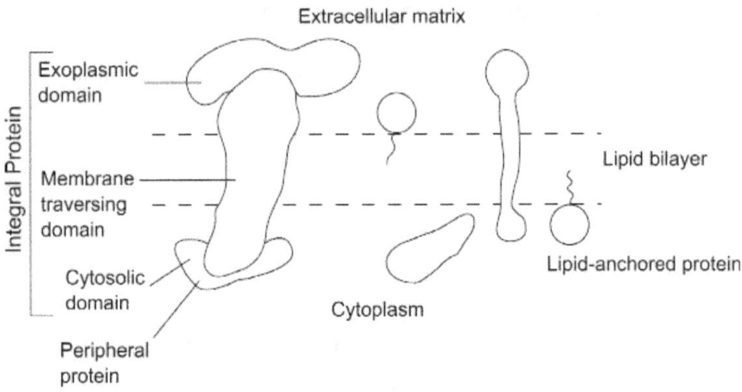

Figure 2.5. Proteins associated with the lipid bilayer.

Integral membrane proteins   They have one or more hydrophobic a helices and hydrophilic regions and extend between cytoplasm and extracellular matrix. These proteins are made up of cytosolic domains, exoplasmic domains and membrane-traversing domains. Of them, the cytosolic and exoplasmic domains have hydrophilic outer surfaces and the membrane traversing domains have many hydrophobic amino acids. The cytosolic and exoplasmic regions interact with aqueous solutions on either side. These proteins are similar to the structure and composition of water-soluble proteins. They always occur in combination with complex sugar groups linked with amino acid side chains (**glycosylation**). The membrane-traversing domain interacts with the hydrocarbon core of phospholipid bilayer and consists of one or more a- helices or many b-sheets. These a helices mostly possess primary and secondary structures. Integrins are different dimeric integral membrane proteins, which act as adhesion receptors in cell-matrix interactions.

Lipid-anchored membrane proteins   These are bound to lipid molecules by covalent bonds in which the hydrophobic carbon of the lipid links the protein to the membrane. These are asymmetrically located with the membrane faces.

Peripheral membrane proteins   These are bound to the membrane either directly with the help of lipid head groups or indirectly with the help of integral membrane proteins. They are found either in the extracellular matrix or in the cytoplasm.

Membrane permeability   Permeability is the fundamental function of living cells to maintain intracellular physiological conditions. The plasma membrane of the cells plays a key role in establishing a difference in the concentration of substances between the intra and extracellular fluids, so as to maintain an appropriate environment for normal cell functioning. The membranes are selectively permeable and act as a barrier for some substances and as a passage for some other substances. This regulatory function of the membranes is called membrane **permeability**. In order to study the mechanism of **membrane** permeability, it is important to know the chemical and molecular organisation of the membranes. The biomembrane is highly asymmetrical formed by lipid bilayer with protein complexes embedded to both the lipid aqueous interfaces. Moreover, the chemical constituents are not uniformly distributed between the inner surface, which is in contact with the cytoplasm, and the outer surface, which is in contact with the surrounding fluid medium.

Transport of water   Water is transported through the plasma membrane by osmosis and diffusion (active transport and passive transport). By endosmosis, water enters into the cell from the surrounding medium whereas water passes out of the cell by exosmosis. The living cells are always bathed in some fluid medium called extracellular fluid. If the extracellular fluid has the same osmotic pressure like that of cytosol (intracellular fluid), then it is called **isotonic solution** in which the cells remain intact. If it has lesser osmotic pressure than cytosol, then it is called **hypotonic solution** in which the cells become swollen. If it has higher osmotic pressure than cytosol, the extracellular fluid is called **hypertonic solution**.

Transport of ions   The ionic concentration of the intracellular fluid differs from that of extracellular medium so that an electrical potential exists across the membrane. The diffusion of ions across the membrane depends on both the concentration gradient and the electrical gradient, because the ions are charged particles. The membrane permeability also involves energy-requiring mechanism called **active transport**, which is coupled with cell respiration. For example, the presence of higher concentrations of potassium ions inside the cells is due to the entry of these ions against concentration gradient thereby requiring energy and the sodium ions are transported to the outside of the cell by the sodium pump which is an active process. The ionic transport occurs through the molecular machinery or ion channel found within the cell membrane. The ion channel is hydrophilic in nature with two functional elements namely a selectivity filter to determine the type of ion to be transported and a gate to regulate the flow of ions. The active transport mechanism also involves carrier molecules. For example, the active transport of Na^+ and K^+ ions depends on the activity levels of ATPases, which are membrane proteins. These enzymes act as complex carrier molecules in the transport of ions as shown in the following equations.

$$E + ATP + Na^+ \rightarrow [Na^+ \ E\sim P] + ADP$$

$$[Na^+ \ E\sim P] + K^+ \rightarrow Na^+ + K^+ + Pi + E$$

$$\uparrow \qquad \downarrow$$

$$out \quad in$$

Here a phosphorylated complex of the enzyme (Na$^+$ E~P) is formed in the presence of Na$^+$ ions on the inner side of the membrane. Then the complex is hydrolysed to form free enzyme and Pi in the presence of K$^+$. In this way Na$^+$ and K$^+$ ions are transported in opposite directions so that electrical neutrality is maintained.

Transport of molecules Permeability of molecules is determined by their chemical nature and thus their transport is highly specific. There are two hypotheses to explain the molecular mechanism of selective transport of molecules across the membrane. According to the **hypothesis of carrier mechanism**, the molecule first binds with the carrier protein at the outer surface of the cell to form carrier-molecule complex, which rotates and liberates the molecule to the cytoplasm of the cell. This theory is not favoured because the possibility of rotation of the carrier molecule inside the membrane is less. According to **fixed pore mechanism** the integral proteins which transverse the membrane act as carriers. This protein, after binding with the molecule to be transported, undergoes structural changes to release the molecule into the cell. This theory is more accepted because the mechanism requires only less energy, which is provided by ATP molecule.

2.4 EXTRACELLULAR MATRIX

The cells of multicellular animals are embedded in a mesh of proteins and polysaccharides called **extracellular matrix** (ECM) which forms the immediate environment of cells. The constituents of extracellular matrix are secreted by cells into the spaces between them. The ECM is responsible for the aggregation of cells into tissues as well as for the transfer of information between the internal and external environment of the cells. The cell-surface receptors or transmembrane adhesion receptors bind the components of ECM and cause adherence of cells to each other. In tissues, the cells adhere directly with each other (**cell-cell adhesion**) by means of specialized integral membrane proteins called **cell-adhesion molecules** (CAMs). On the other hand, the cells adhere indirectly with each other (**cell-matrix adhesion**) by the binding between the receptors in the plasma membrane and the components of ECM. The epithelial and the most organized groups of cells contain specialized extracellular matrix called **basal lamina**, which is a sheet-like mesh underlying or surrounding the cells.

2.4.1 Extracellular Matrix in Epithelial Tissues

The major components of ECM in epithelial tissues include proteoglycans (glycoproteins), collagens (fibre-forming proteins) and multi adhesive matrix proteins, the volume of which varies in different tissues and organs.

Proteoglycans   The ECM of all animal tissues contains large quantity of proteoglycans of different groups. Proteoglycans are glycoproteins containing covalently linked special polysaccharide chains called **glycosaminoglycans** (GAGs) with negative charges. These include various groups of glycoprotein molecules found abundant in the ECM as well as on the cell surface. **Perlecan** is the major secreted proteoglycan in the basal lamina containing a large multidomain core with three or four GAG chains. **Syndecans** are cell surface proteoglycans which help to bind collagens with multiadhesive matrix protein. Therefore, they are useful in attaching the cells to the ECM.

Collagens   In most tissues, collagen is the major component of the extracellular matrix. Though the structure of basal lamina differs in different tissues, it typically consists of four important protein components in the extracellular matrix. Among them, type IV collagen forms the major component of the ECM of the basal lamina. These types of collagens are trimeric molecules having rod-like and globular domains with two-dimensional network. These include more than 20 types of collagens, which take part in the formation of ECM in various tissues. These trimeric proteins are formed by three polypeptide chains called collagen α chains, which may be identical or different. In each strand one of the a chains is twisted into a left-handed helix so that three such strands coil with each other to form a right-handed triple helix held by hydrogen and peptide bonds.

Multiadhesive matrix proteins   **Laminins** are the major multiadhesive matrix proteins in the basal lamina. They are heterotrimeric proteins having different polypeptide chains, which occur in the form of fibrous two-dimensional networks. These are long flexible molecules which bind with various types of collagens, other proteins, polysaccharides, adhesion receptors and extracellular signaling molecules. They also help to bind steroids. **Entactins** possess rod-like domains which link type IV collagens and laminins. In many vertebrate cells, the major multiadhesive matrix protein is **fibronectin**. It is helpful in the adhesion of cells to the extracellular matrix by binding with fibrous collagens, proteoglycans or integrins. These are dimers linked by two disulphide bonds at C-terminals.

2.4.2 Extracellular Matrix in Non-epithelial Tissues

More or less similar molecules are involved in the cell-matrix interactions in non-epithelial tissues. The connective tissue is mostly made up of extracellular matrix containing insoluble protein fibres, proteoglycans, adhesive proteins and non-sulphated hyaluronan. Nearly, 80–90 % collagens found in the connective tissues include type I, II and III collagens and I which are collectively called as **fibrillar collagens** which are differentiated by their ability of coiling to form fibrils. The mechanical properties of the matrix are due to large proteoglycan aggregates, which contain a central hyluronan molecule, linked to the core proteoglycan molecule by non-covalent bonds. Many non-epithelial cells have integrin-containing aggregates, which connect the cells to the ECM.

Dystroglycan forms a large complex with dystrophin (an adapter protein) and links the actin to the surrounding matrix. The neural cell adhesion molecule causes cell adhesions in neural tissues. The connection proteins form transmembrane channels, which connect the cytoplasm of two adjacent cells.

Though the ECM gives mechanical support to the tissues, it also performs several other functions such as strengthening of tendons, tooth or bone, cushioning in cartilages, acting as a reservoir for many extracellular molecules providing a lattice for cellular movements and so on.

PROTEINS

2.5 ORGANISATION OF PROTEIN

Proteins are macromolecules as they are polymers of many amino acids linked by peptide bonds. The monomers of proteins are different amino acids each of which has a central α carbon atom linked to four chemical groups namely an amino group (NH_2), a carboxyl group (COOH), a hydrogen atom (H) and a variable group (R) as side chain as shown the Figure 2.6. Except glycine, the α C is asymmetric in all amino acids so and hence occurs in two mirror-image forms namely dextro (D) and levo (L) isomers (optical isomers) (Figure 2.7). The proteins are found to contain only levo forms of amino acids with few exceptions. The properties of amino acids are determined by their side chains, which differ in size, shape, charge, solubility in water and reactivity. Based on the solubility in water, the amino acids are classified into **hydrophilic amino acids** (basic amino acids, acidic amino acids and polar amino acids with uncharged R groups) and **hydrophobic amino acids** (aliphatic amino acids). The hydrophilic amino acids contain polar side chains found on the surfaces of proteins whereas the hydrophobic amino acids contain nonpolar side chains which aggregate to form water-insoluble cores of proteins. Some hydrophilic amino acids are either positively charged (arginine and lysine) or negatively charged (glutamic acid), thus contributing to the charge of the protein molecules. Some amino acids possess specialized side chains. For example, cysteine has a sulphydroxyl group which is responsible for the formation of covalent disulphide bonds. Glycine has a single hydrogen atom which allows it to fit into tight spaces. Proline has a side chain which forms a ring with nitrogen atom, thus providing rigidity to protein molecules. In a protein molecule, the side chains may be cross-linked through disulphide bonds, which are found in extracellular proteins helping to stabilize the folded structure of proteins.

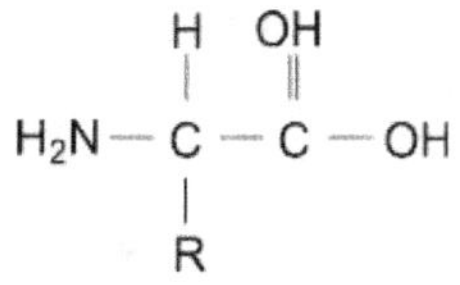

Figure 2.6. An amino acid monomer.

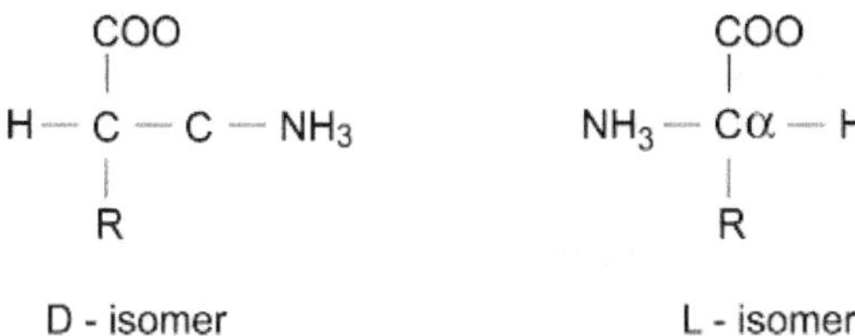

D - isomer

L - isomer

Figure 2.7. D and L isomers of an amino acid.

2.5.1 Primary Structure

In biological systems, proteins are the most abundant macromolecules and have specific conformation to suit their functions. In a protein molecule, the amino acids are covalently linked in such a way that the amino group of one amino acid is linked with the carboxyl group of another amino acid. This type of linkage is called **peptide linkage** (–C–N–). The primary structure of proteins represents the linear sequence of amino acids in which the peptide linkages are formed between amino acids in the form of a long linear chain. One end of the chain has a free amino group (N-terminal) and the other end has free carboxylic group (C-terminal). For example, insulin is a protein molecule which consists of two polypeptide chains namely A chain with 21 amino acids and B chain with 30 amino acids. The two chains are held together by two disulphide (–S–S–) groups at positions of 7-7 and 20-19. The chain A has glycine as N-terminal and aspartic acid as C-terminal. The chain B has phenylalanine and alanine respectively as C- and N- terminals. In addition, the chain A has an interchain –S–S– between sixth and eleventh amino acids. The structure of insulin is shown in the Figure 2.8.

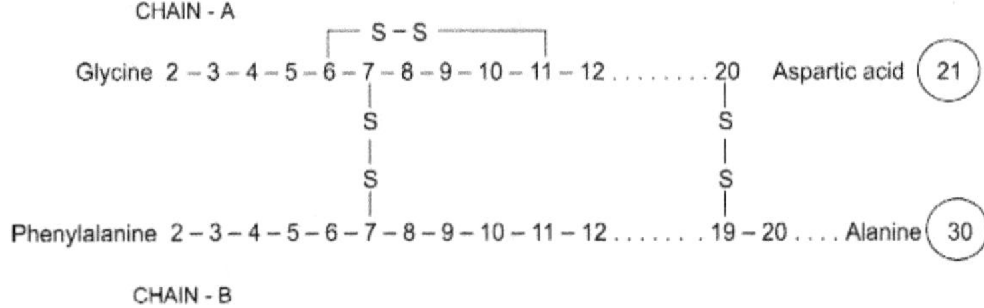

Figure 2.8. Structure of insulin.

2.5.2 Secondary Structure

The secondary structure of proteins represents folding of some parts of a polypeptide chain resulting in various spatial arrangements of amino acids. A polypeptide chain exhibits random coiling in the absence of the stabilizing non-covalent interactions. When stabilizing hydrogen bonds occur, the protein molecule may fold to from an α helix or β sheet.

α-Helix   It is highly rigid, rod-like and most common. The best example for α helix is keratin which is a fibrous protein. The participation of CO and NH groups in the formation of hydrogen bonds is shown in Figure 2.9. Here the hydrogen bonds formed are intramolecular and the entire chain takes the shape of a helix.

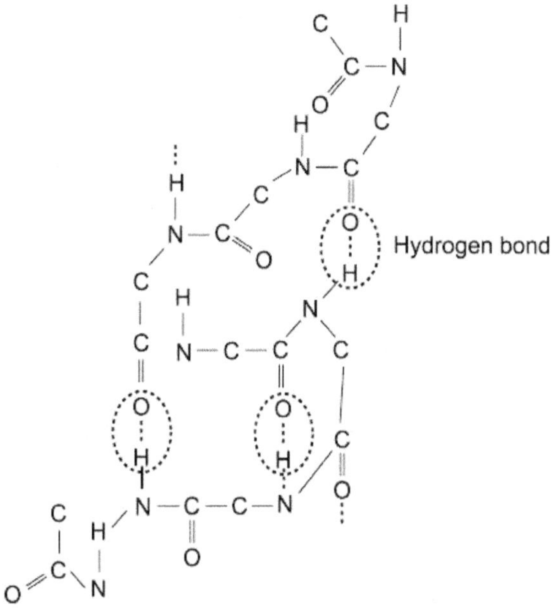

Figure 2.9. Structure of keratin.

β-*pleated sheet*   The protein fibroin exhibits another type of secondary structure consisting of a number of adjacent straight polypeptide chains which run parallel in the same direction or in opposite direction. The chains are held together by hydrogen bonds between CO and NH groups of one chain and NH and CO groups of adjacent chains as shown in the Figure 2.10.

The structure of β-pleated sheet is very stable and rigid because of the participation of CO and NH groups in the formation of hydrogen bonds. Collagen is also a fibrous protein in which three polypeptide chains coil around each other to form a triple helix held together by hydrogen bonds.

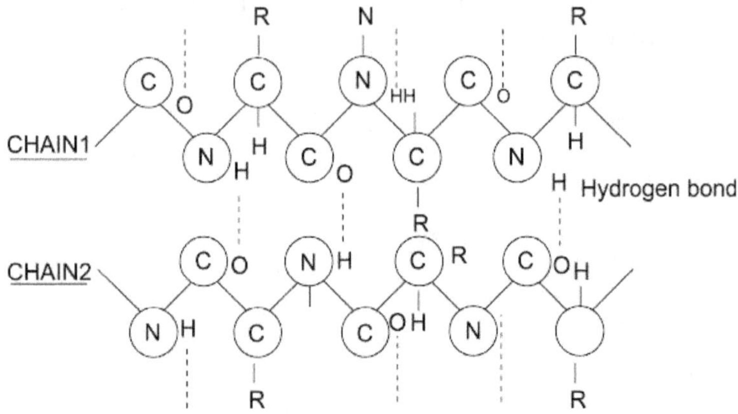

Figure 2.10. Structure of fibroin.

Many such triple helices form fibrils in which the chains lie parallel to each other, linked by covalent bonds.

2.5.3 Tertiary Structure

The three-dimensional arrangement of amino acid residues in a polypeptide chain represents the tertiary structure of proteins in which a polypeptide chain folds itself by bending with the resultant compact globular shape. The tertiary structure of proteins involves the hydrophobic interactions between nonpolar side chains and hydrogen bonds between polar side chains as well as disulphide and ionic bonds as in Figure 2.11. As a result, these interactions transform the secondary structure of a protein molecule to a compact form. The tertiary structure of proteins is due to particular combinations of secondary structures called **motifs** or **folds**. For example, the protein myoglobin has a primary structure in which the single polypeptide chain contains 153 amino acid residues. It occurs mostly in the form of α helices due to the presence of hydrogen bonds between CO and NH groups of amino acid residues. By folding, it becomes globular in shape due to the formation of various bonds and hydrophobic interactions. The diagrammatic three-dimensional structure of myoglobin is shown in Figure 2.12.

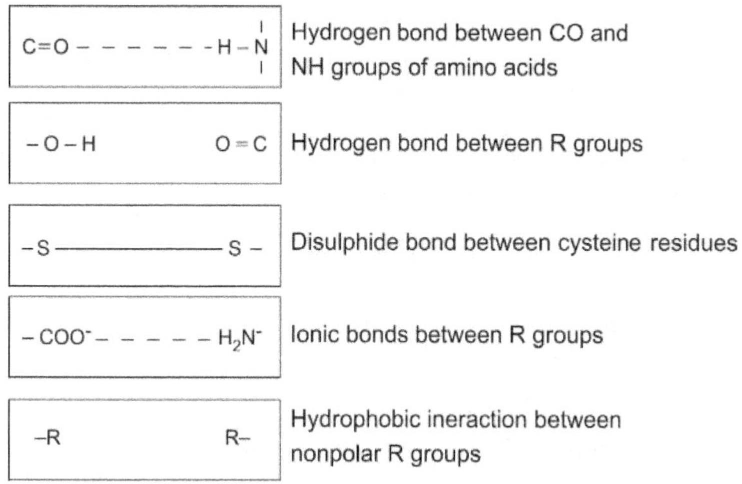

Figure 2.11. Main linkages in tertiary structure of proteins.

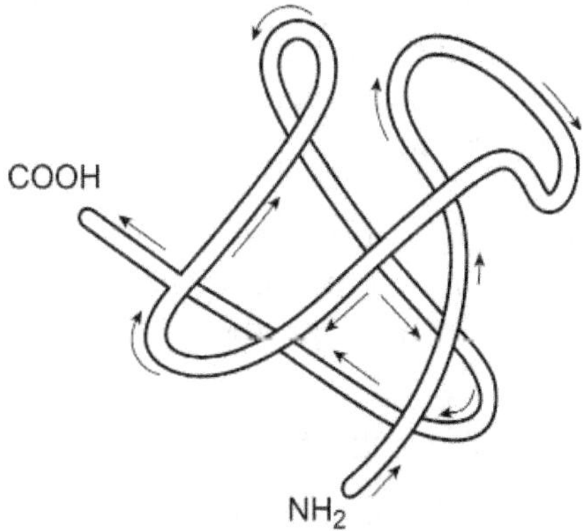

Figure 2.12. Diagrammatic representation of the tertiary structure of myoglobin molecule.

2.5.4 Quaternary Structure

Many proteins are highly complex as they contain many polypeptide chains held together by different bonds and hydrophobic interactions. Thus two or more polypeptide chains constitute multimeric proteins, the number and relative positions of the subunits of which are explained by the quaternary structure. For example, **haemagglutinin** is a trimer having three similar subunits linked by non-covalent bonds. **Haemoglobin** has four polypeptide chains namely two α chains and two β chains. Each α chain contains 141 amino acid residues whereas each β chain contains 146 amino acid residues. As in other globular proteins, the hydrophobic side chains face the centre of the molecule and the hydrophilic side chains outwards.

Thus the primary structure of proteins explains the sequence of amino acids which are linked by peptide bonds. The secondary structure is due to folding of the peptide chain into a helix by hydrogen bonds. The tertiary structure is due to the further folding to form globular or any other conformation due to disulphide bonds and van der Waal's and hydrophobic forces. In short, the proteins with a single peptide chain exhibit primary, secondary and tertiary structures whereas the proteins as polymers exhibit quaternary structure. Generally, the protein molecules with high molecular weight possess quaternary structures. The compact and locally folded region of the tertiary structure of the protein forms a domain and many proteins are made up of more than one domain. These domains run through the whole molecule.

2.5.5 Cellular Aspects of Protein Folding

The proteins synthesized by ribosomes may remain in the cytosol or may be transported to their destinations. The transportation of proteins to different regions of a cell is called **protein targeting** or **protein sorting** involving two important processes. In the first process, the proteins are transported to the membranes of intracellular organelles where the proteins combine with the lipid bilayer of the membrane or reach the interior of the organelles. In the second process, the proteins are transported to the membrane of the endoplasmic reticulum for secretory purposes. Before being transported to their final target sites, the membrane and secretory proteins undergo four important changes namely glycosylation in the endoplasmic reticulum and golgi, formation of disulphide bonds, folding of polypeptide chains and proteolytic breakdown in endoplasmic reticulum. Only the assembled and properly folded proteins are transported whereas the unfolded or partially folded proteins are retained in the rough-surfaced endoplasmic reticulum whereas the unassembled or misfolded proteins are readily degraded. A fast and more efficient folding of proteins is necessary for cellular functions and is achieved by a group of proteins in cellular compartments. These proteins are called **chaperones** which are of two types namely **molecular chaperones** which prevent the aggregation and degradation of proteins by stabilizing unfolded or partially folded molecules and **chaperonians** which are directly involved in the folding of proteins.

The tertiary and quaternary structures of many proteins are stabilized by the presence of disulphide bonds which are found only in secretory and exoplasmic membrane proteins. In cytosolic proteins, the disulphide bonds are absent and they are stabilized by other interactions. The proteins which are transported from the cytosol to the mitochondria, contain **matrix-targeting sequences** located at the end terminal of amino acids. These amino acids are basic in nature, hydroxylated, positively charged and hydrophobic. They have α helical conformation with positively charged amino acids on one side of the helix and hydrophobic amino acids on the other side. In general, the unfolded proteins are imported into the mitochondria and some of them fold into their final conformation with the help of chaperonin and ATP. Unlike mitochondria, only folded proteins are transported across the membrane of peroxysomes with the help of ATP molecules.

2.6 INTERACTION OF PROTEINS

The normal functioning of a living cell is not only due to gene expression but also due to different proteins and their associations with each other. The activities of many proteins depend on their interaction with each other or with small intracellular signaling molecules.

2.6.1 Protein–Protein Interactions

The ions and molecules are in constant collision both inside and outside the cells. When two molecules come together, they are repelled because of the presence of weak non-covalent bonds. However, the occurrence of complementary regions enables the molecules to form a large number of non-covalent bonds so that they link with each other. The different specific weak bonds or non-covalent interactions between complementary regions are able to bind two protein molecules as in Figure 2.13. These interactions within a protein molecule cause it to fold into three dimensional conformations.

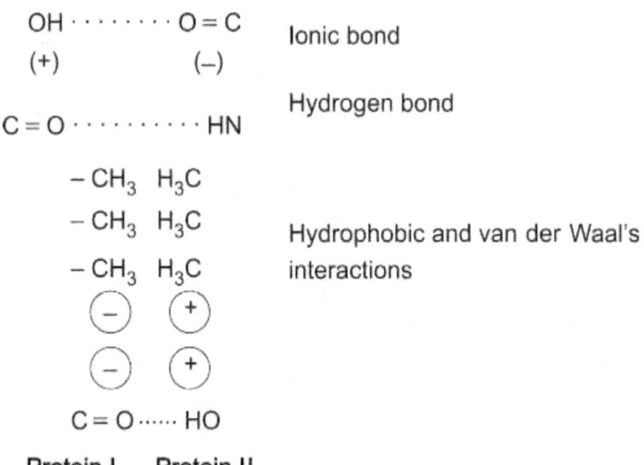

Figure 2.13. Binding of proteins by non–covalent interactions.

Thus many proteins in a cell interact with each other harmoniously so that their activities are regulated. When the metabolites are required, the synthetic reactions proceed whereas they are broken down when they are degraded. In this way, the regulatory mechanism found inside the cell alters the conformation of a protein at its site of action.

Moreover, segregation of proteins to definite compartments like mitochondria, nucleus and lysosome is also important. The sequence of a secretory protein targeted to endoplasmic reticulum consists of a segment of hydrophobic amino acids found at the N–terminals. Here cytosolic ribonucleoproteins act as **signal-recognition particles** (SRP) and bind with **endoplasmic reticulum-signal sequence** on the protein, thus targeting to the endoplasmic reticulum. The protein molecules are **polyampholytes** because they have many acidic and basic groups thus constituting macroions. Therefore, the proteins carry electrostatic forces of attraction or repulsion depending on the solution pH, which determines their solubility.

2.6.2 Protein–Ligand Interactions

The functions of almost all proteins depend on their ability to bind with other molecules namely **ligands**, A ligand may be an activator, inhibitor or a substrate or all the three. This is because the ligand-binding sites on proteins and the corresponding ligands are complementary with each other. The binding ability of a protein molecule depends on the strength of binding. Any alteration in the tertiary or quaternary structure of proteins induced by a ligand is called **allostery**. Many ligands such as oxygen, cAMP and ATP cause allosteric change in target proteins. In addition, other allosteric ligands such as calcium ions and GTP also act on proteins and regulate cellular functions. When a protein molecule binds with several molecules of a ligand, the binding of one ligand molecule affects the binding of subsequent ligands. This type forms graded binding and is called **cooperativity** in which the proteins respond efficiently even at small changes of ligand concentration. Thus in protein-ligand interaction the enhanced binding shows positive cooperativity, whereas the inhibited binding denotes negative cooperativity. For instance, haemoglobin is an example for positive cooperativity. It has four haem subunits in which the binding of oxygen to one of the subunits induces a conformational change. This spreads to other subunits to bind additional oxygen molecule.

In living organisms, the extracellular signalling molecules act as ligands and control the metabolic processes inside the cells. These molecules produce specific responses only in target cells, which possess receptors for binding with ligands. The plasma membrane contains a variety of receptor proteins to bind with ligands. A ligand binds to a complementary site on the membrane receptor. This binding of extracellular signaling molecule to cell surface receptors causes a conformational change in the protein structure inducing cellular responses. These extracellular signaling molecules include secreted proteins, peptides, small lipophilic substances like steroid hormones, small hydrophilic substances derived from amino acids like epinephrine, gases like nitrous oxide and physical stimuli like light. The signals can act on the signaling cell itself (**autocrine**) or on nearby cells (**paracrine**) or on distant cells (**endocrine**). The binding of ligands with receptors is due to non-covalent interactions between a ligand and specific amino acid in the receptor proteins. The responses to the external signal occur even when a fraction of receptor molecules are occupied by ligands. The binding of ligand to cell surface receptor may increase or decrease the level of calcium ions, cAMP, etc. The signal binding proteins include monomeric and trimeric proteins as well as protein kinases and phosphatases.

Protein kinase A is an enzyme occurring as an inactive tetrameric protein with two catalytic subunits and two regulatory subunits. Each regulatory subunit has a pseudosubstrate sequence which binds with the catalytic subunit. Here the regulatory subunit inactivates the catalytic subunit by preventing the substrate binding. The presence of cyclic AMP causes a conformational change in the pseudosubstrate sequence so that the inactive tetramer dissociates into two monomeric catalytic subunits and a dimeric subunits as in Figure 2.14.

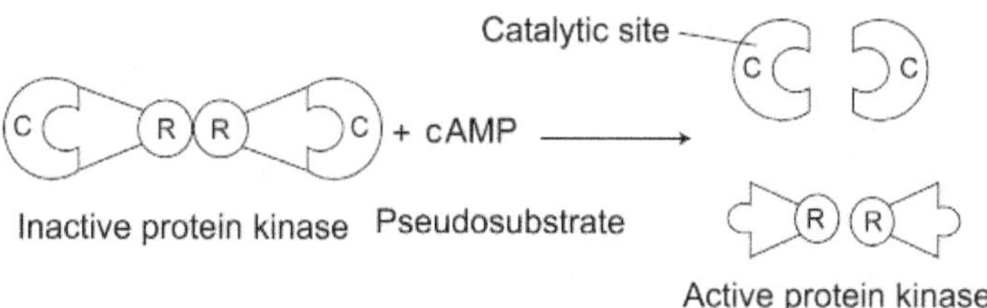

Figure 2.14. Schematic representation of activation of protein-kinase through ligand binding.

In cells, the concentration of free calcium ions is kept at a very low level but the concentration can be increased by the release of calcium ions from endoplasmic reticulum or from extracellular environment. These calcium ions bind with calcium-binding proteins called **calmodulin** in the cytosol. These proteins exist either in the form of an individual monomeric protein or subunits of a multimeric protein with four calcium binding sites. The binding of calcium ions to calmodulin produces a conformational change so that it binds with target proteins thus influencing their activity.

Guanine nucleotide-binding proteins form another type of intracellular switch proteins which occur in two forms namely an active form which binds with GTP and regulates the activity of specific target protein and an inactive form which binds with GDP. These proteins hydrolyse bound GTP to GDP, thus converting it to an inactive form. However, the regeneration of GTP occurs very slowly

Phosphorylation alters the charge of proteins thus causing conformational changes. This would in turn alter the capacity of the protein to bind with ligands leading to altered activity level.

2.7 SEQUENCING OF PROTEINS

Protein sequencing refers to the determination of the primary structure of a protein and can be done by a number of ways. The proportion of different amino acids in a protein molecule can be determined by chromatographic separation of the hydrolysed protein. **Edman degradation method** is the classic technique to determine the amino acid sequence of a protein. **X-ray crystallography** is used to study three-dimensional structure of proteins. The structure of proteins which are difficult to crystallize, can be studied by **cryoelectron microscopy**. Three-dimensional structure of small proteins containing about 200 amino acids can be studied using **NMR spectroscopy**.

2.7.1 Determination of Amino Acid Terminals

To determine the amino terminals of amino acids, a protein is first treated with Sanger reagent (1-fluoro-2-4-dinitrobenzene) and completely hydrolysed. The extracted dinitrophenyl amino acid indicates the amino terminal of the protein. The second and third amino acids in the amino terminal can be determined by treating the extract in Edman reagent (phenyl isothiocyanate).

To determine the carboxyl terminals of a protein molecule, the protein is first treated with hydrazine and then hydrolysed so that the amino acids at the carboxyl terminals occur as hydrazide. The carboxyl terminal extract is then treated with the enzyme carboxypeptidase to separate the amino acids. When the number of amino acids at each terminal of the protein molecule is determined, the protein is broken down into small peptides, the sequence of which can be determined.

2.7.2 Protein Sequencing through Determination of Gene Sequences

A particular gene transcribes the primary structure of each protein. When the code which relates DNA sequence to protein sequence is known, the determination of nucleotide sequence will give the corresponding protein sequence. However, this method is helpful to know the protein sequence of a protein synthesized as such. It does not provide information regarding modifications of amino acids or the occurrence of intramolecular bonds.

2.7.3 Edman Degradation Method

By this method, the protein is treated with phenyl isothiocyanate in the presence of an alkali. Now the terminal amino group of the polypeptide chain yields a phenylthiocarbonyl derivative of the peptide chain. When this derivative is treated with a strong anhydrous acid, the peptide bonds between the residues 1 and 2 are degraded. As a result, the derivative of the N-terminal residue rearranges to give phenylthiohydantoin (PTH) derivative of the amino acid. The second residue is determined by repeating the whole reaction. In this way, a long polypeptide chain can be sequenced from the N-terminal end.

2.7.4 Sequenators

At present, the direct protein sequencing is accomplished by using instruments called sequenators, which are capable of carrying out entire reactions of the Edman method. The instrument saves the PTH derivative of each amino acid residue of a polypeptide chain in separate tubes for many cycles. The PTH derivatives are identified by high performance liquid chromatography (HPLC). A number of instruments are put to use to sequence peptides using the Edman technique.

Solid-phase liquid-pulse sequenator   The reaction vessel of this sequenator is shown in Figure 2.15. A thin polymer membrane is held in the reaction vessel to which the protein molecules are covalently attached through its C–terminal. The reagents are added to the upper chamber of reaction vessel in which the reactions occur and phenylthiohydantoin derivatives are automatically collected and analysed. In **gas-phase sequenator**, the reagents are added in gaseous state so that the losses are minimized and the analysis is more rapid.

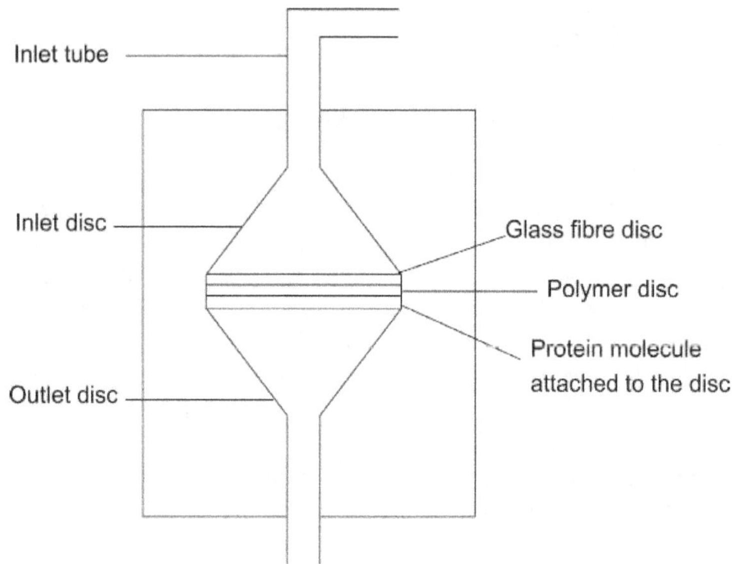

Figure 2.15. Reaction vessel of solid-phase sequenator.

BLAST algorithm   Proteins with same functions contain similar amino acid sequences which correspond to the functional domains in the three-dimensional structure of proteins. These sequence similarities for the function of encoded protein can be obtained by comparing the amino acid sequence of protein encoded by a gene with the sequences of proteins of known function. The related protein exhibits more sequence similarity than the genes encoding them and thus the protein sequences are compared. The computer programme employed for this comparison is called BLAST (Basic Local Alignment Search Tool). The BLAST algorithm divides the new protein sequence into shorter segments and searches for particular matches to any of the stored sequences. When a particular match is identified for a segment, it further searches for the region of similarity. Then the programme ranks the matches depending on their P-values, which are the measures of the probability of the degree of similarity between two protein sequences by chance. If the P-value is lower, then the sequence similarity between the two sequences is greater.

2.8 PROTEIN DENATURATION

Proteins are macromolecules formed by amino acids with acidic and basic groups linked together by peptide bonds. In other words, amino acids combine together by peptide bonds to form a protein molecule which typically folds into a particular shape. Structurally proteins exhibit four configurations namely primary, secondary, tertiary and quaternary structures. These configurations are due to ionic, disulphide and hydrogen bonds as well as due to hydrophobic interactions.

2.8.1 Bonds in Protein Molecules

Peptide bond　It is a covalent bond formed by the reaction between amino group of one amino acid and the carboxyl

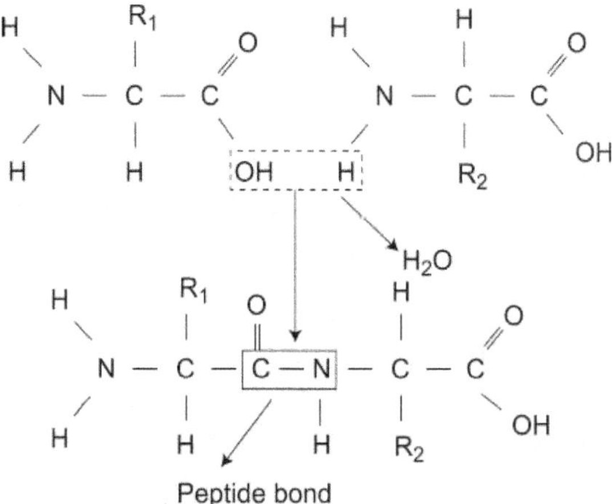

Figure 2.16. Formation of a dipeptide bond.

group of another amino acid with the elimination of a water molecule as in Figure 2.16. Many amino acids are thus combined to form a polypeptide chain.

Ionic bond   In a protein molecule, the acidic and basic R groups occur in an ionized or charged state at certain pH. As the acidic R groups are negatively charged and the basic R groups are positively charged, they attract each other forming ionic bonds as in the Figure 2.17. The ionic bond is much weaker than the covalent bond that even changing the pH of the medium can break this bond.

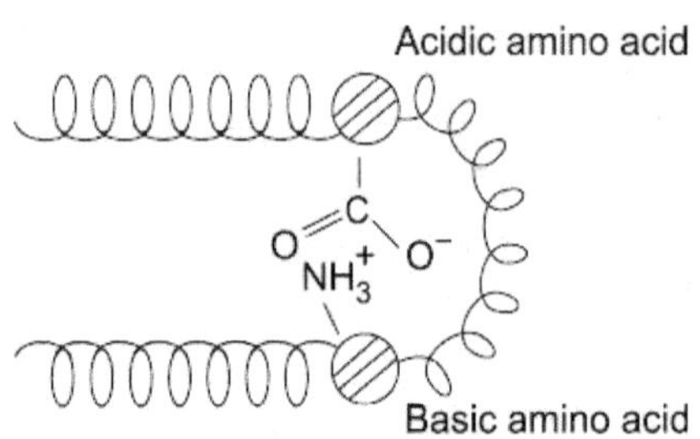

Figure 2.17. Formation of ionic bond (Diagrammatic).

Disulphide bond   Cysteine contains a sulphydryl group (–SH) in its R region. When two cysteine molecules combine together, –SH groups are oxidized so that a disulphide bond is formed as in Figure 2.18. This bond is strong and cannot be easily broken.

Hydrogen bond   In a water molecule, one part is slightly positive and the other part is slightly negative (dipole). This is because the oxygen atom has greater electron-attracting power than the hydrogen atoms. As the electrons are negatively charged, the oxygen atom acquires a slightly negative charge as in Figure 2.19. This hydrogen bond is weak and less strong than the ionic and covalent bonds. In a polypeptide chain, the electronegative oxygen or nitrogen atom may attract hydrogen.

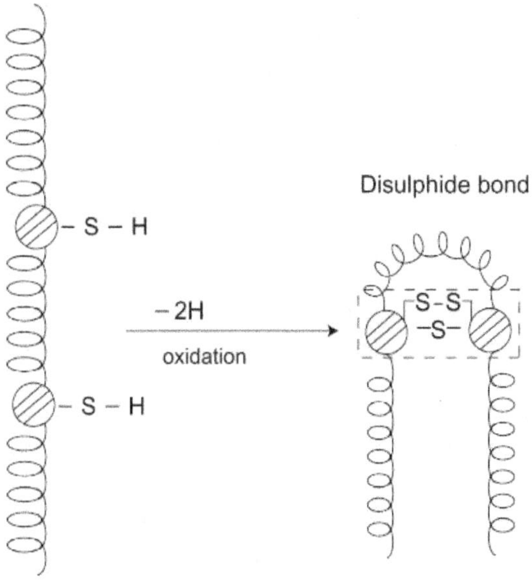

Figure 2.18. Formation of disulphide bond (Diagrammatic).

Figure 2.19. Formation of hydrogen bond between water molecules.

Hydrophobic interaction As the nonpolar molecules are insoluble in water, they form aggregates due to the hydrophobic interaction because the nonpolar molecules cannot form hydrogen bonds.

Protein denaturation The disruption of the protein structure by a denaturing agent and resulting in alterations of its physiochemical and biological characteristics is called protein denaturation. In other words, the loss of specific three-dimensional structure of a protein molecule refers to denaturation, which may be temporary or permanent.

However, the amino acid sequence of the protein molecule is not affected. During denaturation, hydrogen and ionic bonds as well as hydrophobic interactions are broken in a protein molecule whereas the peptide and disulphide bonds are not affected. Thus denaturation of proteins is due to the breakage of hydrogen bonds and nonpolar hydrophobic interactions between amino acid side chains resulting in the unfolding of the coiling of the protein molecule.

Characteristics of denatured proteins   The denaturation may result in the crystallization of proteins with decreased solubility and increased digestibility. The protein molecules become unfolded and uncoiled due to rupture of the cross linkages. In severe denaturation, the proteins are precipitated and made insoluble which are irreversible whereas, the altered protein structure is restored in mild denaturation. When a rigid protein molecule is unfolded into a flexible one, the viscosity is increased. For example, when protein is treated with guanidium chloride, which converts the compact configuration of protein into coil configuration, the viscosity of the protein solution is increased. On the other hand, the denaturation of fibrous proteins decreases the viscosity on treatment with guanidium chloride. This shows that chemical treatment causes the breakdown of the covalent bonds or their formation so that a protein molecule changes its shape with altered viscosity. The denatured proteins are capable of dissociating into smaller particles. During denaturation, the number of reactive – SH– and –S–S– groups increase in protein molecules which have lesser capacity to combine with water. The denatured protein enzymes are inactive. Moreover, the oxyphoric capacity of denatured haemoglobin molecule is found to be lost. The altered structure of protein would result in the destruction of the property of immunoglobulins.

Denaturing agents   Protein denaturation is brought out by many physical agents such as heat, high pressure, UV, infrared and X-rays and ultrasonic sound and chemical agents such as acids, alkalis, heavy metals, urea, alcohol, acetone, detergents, etc. The ultraviolet or infra-red rays cause heavy vibrations in the atoms of protein molecules so that the weak hydrogen and ionic bonds are disrupted resulting in the coagulation of proteins. Similarly strong acids, alkalis and highly concentrated salt solutions interfere with and disrupt the ionic bonds of protein molecule. The cations of heavy metals combine with the negatively charged carboxyl groups of proteins and break the ionic bonds. The organic solvents and detergents affect the hydrophobic interaction by forming bonds with hydrophobic groups of protein molecules. This will cause disruption of hydrogen bonds.

CARBOHYDRATES

2.9 ORGANISATION OF CARBOHYDRATES

Carbohydrates are organic substances formed by carbon, hydrogen and oxygen. They include three major classes namely monosaccharides, disaccharides and polysaccharides. Among them, the monosaccharides and disaccharides can be considered as sugars because they possess sweetness and some common properties. The monosaccharides are formed by single sugar molecules with the general formula of $C_6H_{12}O_6$. Based on the number of carbon atoms, they are classified into trioses (^3C), tetroses (^4C), pentoses (^5C), hexoses (^6C) and so on. Disaccharides are formed by the combination of two monosaccharides especially hexoses through condensation with the removal of water. The general formula of disaccharides is $C_{12}H_{22}O_{11}$. Polysaccharides are formed by the condensation of many monosaccharides through glycosidic bonds, i.e., they are polymers of monosaccharides, forming giant molecules.

2.9.1 Monosaccharides

Monosaccharides occur either in the form of a straight open chain or in the form of a ring due to the bond angles between carbon atoms. Experimental evidences show that the monosaccharides exhibit two types of rings. For example, in glucose (hexose sugar), the first carbon atom combines with oxygen atom on carbon atom number 5 to give a six-membered ring (**pyranose form**). Similarly in ribose (**pentose sugar**), the first carbon atom combines with the oxygen atom on the 4th carbon atom to give a five-membered ring (**furanose form**) as shown in the Figure 2.20. However, the six-membered ring is

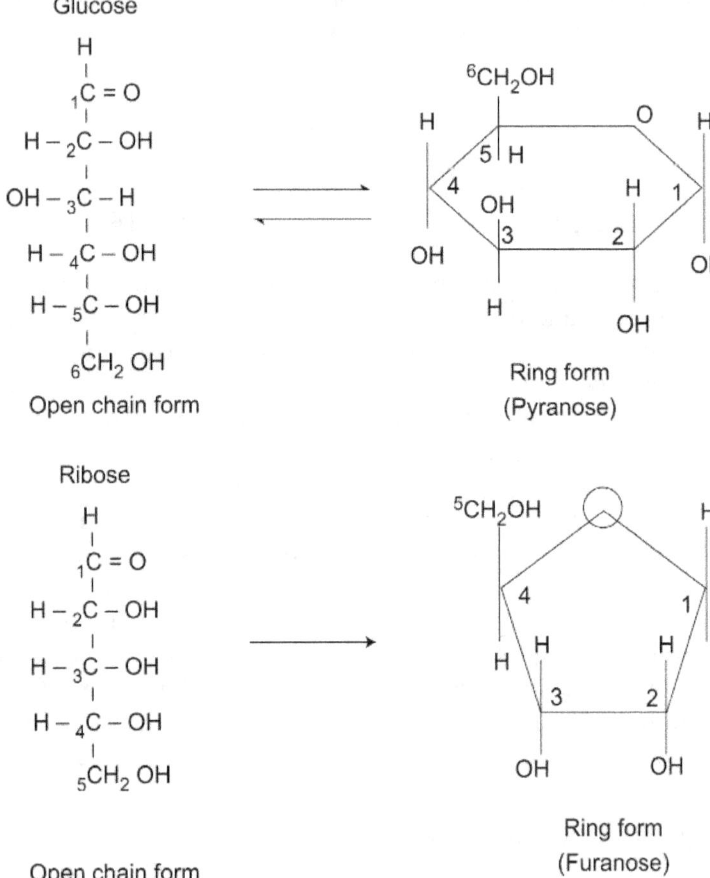

Figure 2.20. Structure of glucose and ribose as open chain and ring forms.

more stable and so glucose occurs predominantly in the pyranose form. In a monosaccharide, when the hydroxyl group on carbon atom 1 projects below the ring, it is known as α ring. If the hydroxyl group projects above the ring, it is called β ring as shown in the Figure 2.21.

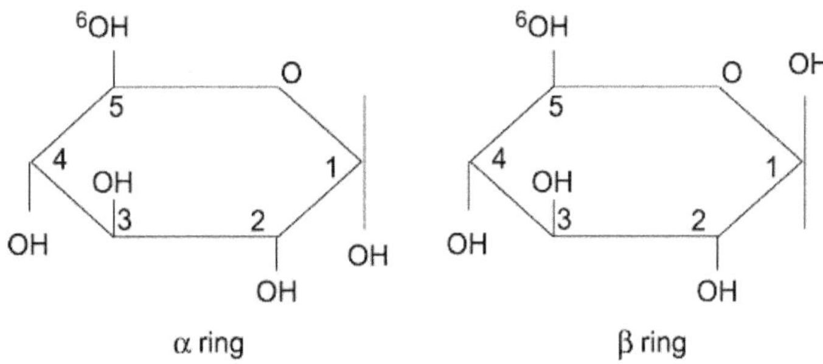

Figure 2.21. Alpha and beta isomers of glucose.

Such molecules with same chemical formula but with different structures are called **isomers**. In glucose solution, the molecules can change from the open chain form to either type of rings or vice versa. In an equilibrium state, these two forms occur in constant proportions. Moreover, a six-membered ring can have either A form or a chair form as in the Figure 2.22. From the energy point of view, a six-membered glucose ring exhibits a chair form in which the covalent bonds

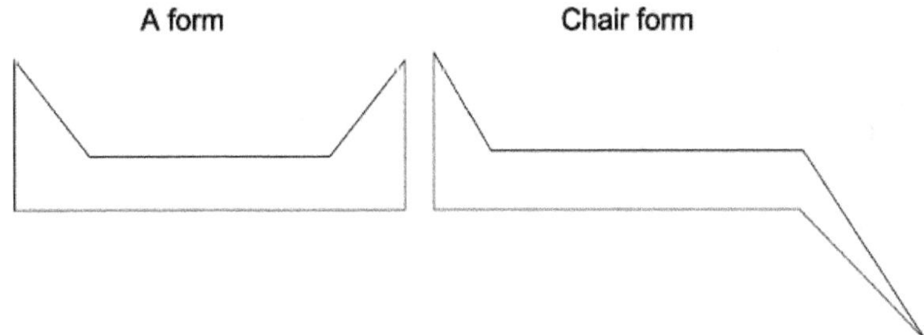

Figure 2.22. Two forms of a six-membered ring (Diagrammatic).

of carbon atoms are either directed upwards and downwards (polar ends) or laterally (axial or equatorial). As in the Figure 2.23, there is difference in the conformation of two OH groups which form part of –CH OH, one being axial and the other polar.

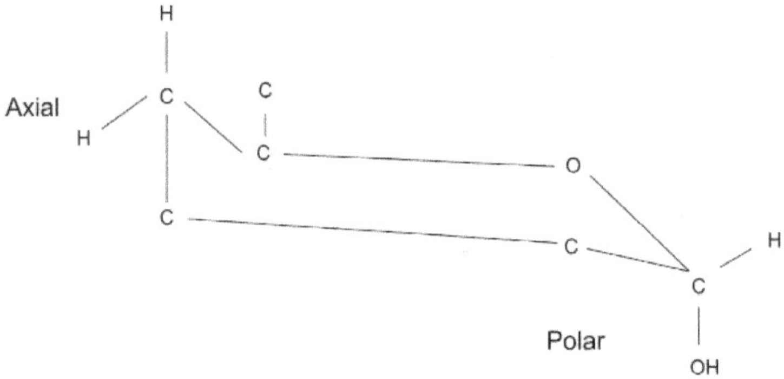

Figure 2.23. Conformation of OH groups in a glucose molecule.

2.9.2 Disaccharides

Two monosaccharides combine with each other to form a disaccharide through the formation of glycosidic bond between carbon atoms 1 and 4 of the neighbouring disaccharides. The monosaccharides which link to form di- or polysaccharides are called **residues**. For example, maltose has two glucose residues as in Figure 2.24. In disaccharides like sucrose, the ring-oxygen atom is a hydrogen bond acceptor and each hydroxyl group is associated with two hydrogen bonds one as a donor and the other as an acceptor. It has two intermolecular hydrogen bonds

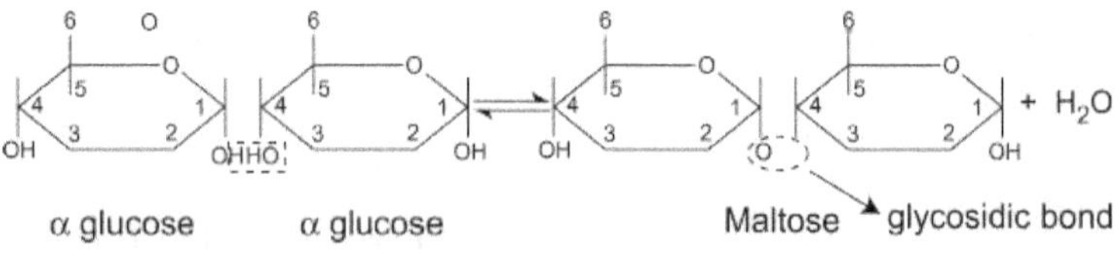

Figure 2.24. Formation of maltose through glycosidic bond.

involving two hydroxy-methyl groups of furanose residues and the ring O(5) and the hydroxyl O(2) of pyranose residues as shown in Figure 2.25.

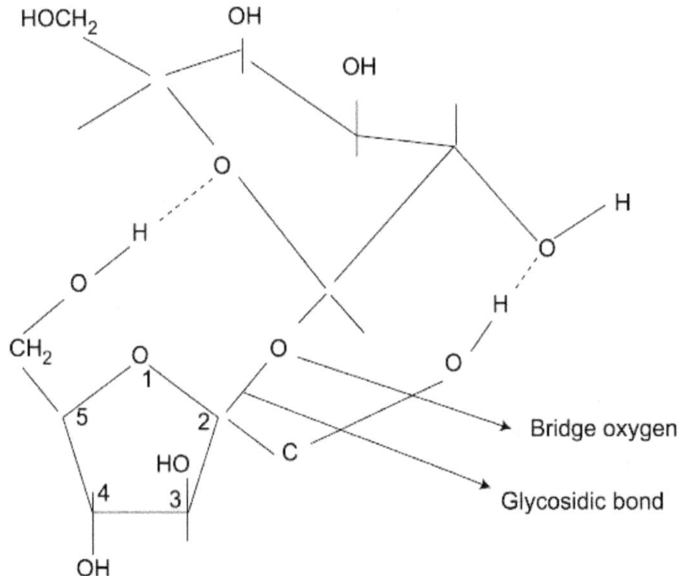

Figure 2.25. Ring structure of sucrose molecule.

2.9.3 Polysaccharides

The polysaccharides are formed by a few to hundreds of monosaccharides through covalent bonds between carbon and water molecules and contain hydroxyl groups. For example, **starch** is a polymer of α-glucose and occurs as amylose and amylopectin. **Amylose** consists of several glucose residues attached by 1, 4 glycosidic bonds as a straight chain as shown in Figure 2.26.

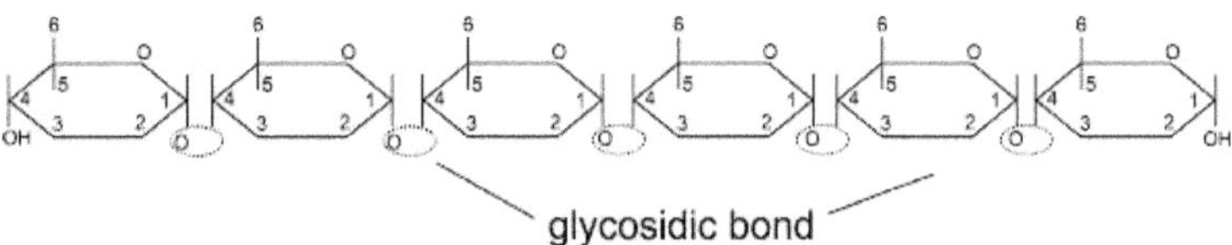

Figure 2.26. Straight chain of amylose.

The glycosidic bonds in amylose cause helical coiling of the straight chain to have a more compact shape. Similarly, **amylopectin** occurs in a compact form with many branches (Figure 2.27) formed by 1,6 glycosidic bonds.

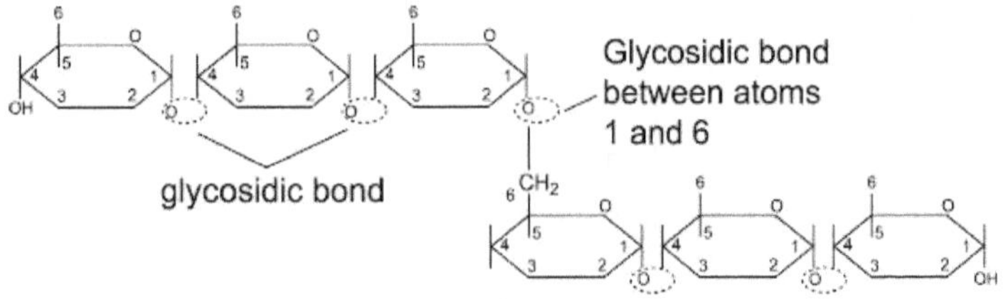

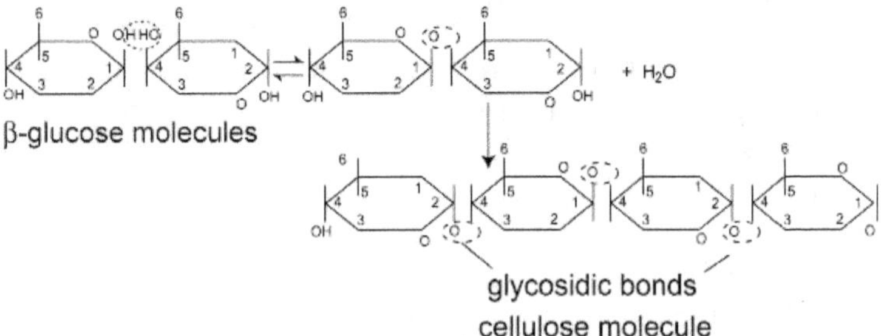

Figure 2.27. Formation of a branch by glycosidic bond in amylopectin.

Glycogen is also made up of many α-glucose molecules and is structurally very similar to amylopectin but with more branching. But **cellulose** is a polymer of β-glucose unlike glycogen and starch. In β-glucose, the –OH group of carbon atom 1 projects below the ring and the – OH group on carbon atom 4 projects above the ring. Therefore, when two β-glucose molecules link together, one of the molecules is rotated through 180° so that the –OH group on carbon atom 1 lies along with the –OH group on carbon atom 4. That is why cellulose has an entirely different structure when compared to the structure of glycogen and starch (Figure 2.28).

Figure 2.28. Formation of cellulose.

Generally, cellulose occurs in the form of fibrils which are grouped into microfibrils. These fibrils are held together by extensive hydrogen bonding as in Figure 2.29.

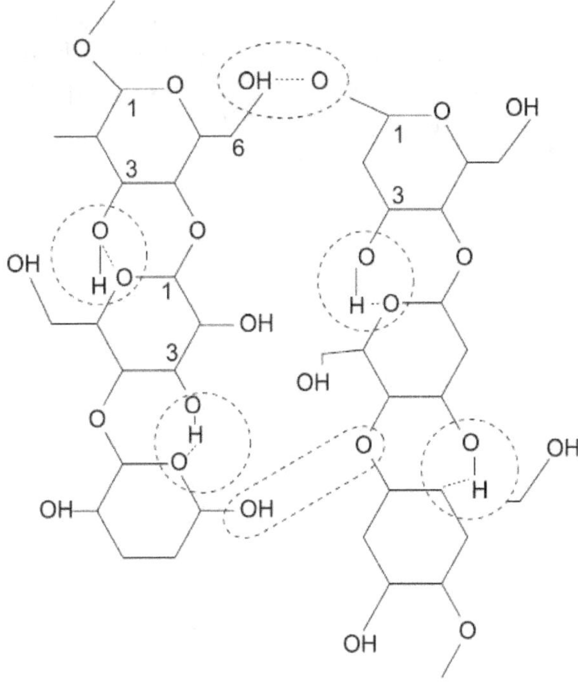

Figure 2.29. Hydrogen bonding in cellulose.

Each cellulose chain has a zigzag conformation with successive glucose residues which are rotated through 180° with respect to each other permitting a H bond between –OH group at carbon atom 3 of one glucose residue with ring O atom of next glucose residue. This impedes the rotation of the adjacent glucose residues around the glycosidic bond and produces a stiff band-like molecule with chain conformation.

2.9.4 Glycoproteins

The oligosaccharides are the simplest disaccharides made up of two residues such as sucrose, lactose and trehalose.

The oligosaccharides (**glycans**) and proteins can be linked to form glycoproteins in two groups namely N-linked glycans and O-linked glycans. N-linked glycans are attached through N-acetylglucosamine or N-acetylgalactosamine to the side chain amino group in an asparagine residue. O-linked glycans are attached by an O-glycosidic bond between N-acetylgalactosamine and the hydroxyl group of a threonine or serine residue as in Figure 2.30.

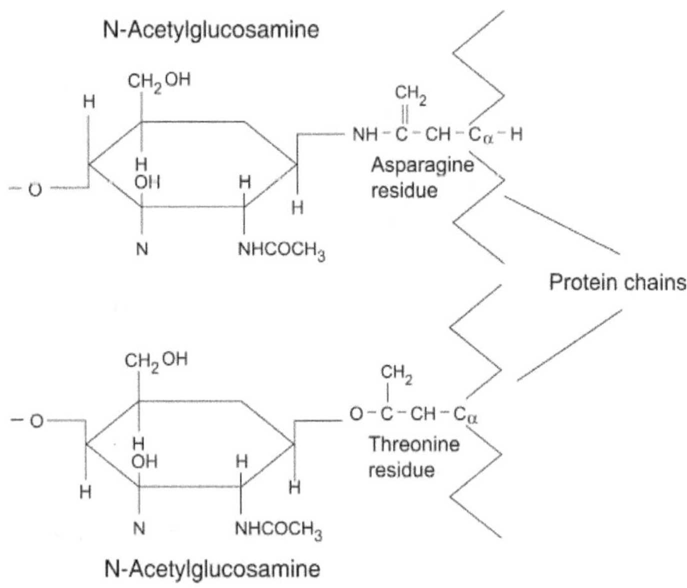

Figure 2.30. Linkage of oligosaccharides with proteins.

Many transmembrane proteins are glycoproteins in which proteins are linked with carbohydrate chain by covalent linkages to serine, threonine or asparagine of the polypeptide chain. **Proteoglycans** are glycoproteins with covalently linked polysaccharide chains called **glycosaminoglycans** (GAGs) which are long, linear polymers of specific repeating disaccharides bearing negative charges. The function of proteoglycans is determined by the arrangement of sugar residues in them and the modification of specific sugars in the chain.

LIPIDS

2.10 ORGANISATION OF LIPIDS

Lipids are not polymers but are small molecules held together by non-covalent bonds. In general, lipids have a polar, hydrophilic head linked to a nonpolar hydrophobic hydrocarbon tail. They are amphipathic molecules forming the surface membrane in living cells, thus playing an important role as major constituents of biological membranes. The simplest lipids are the fatty acids from which many complex lipids are formed. There are four major groups of membrane-forming lipids namely glycerophospholipids, sphingolipids, glycosphingolipids and glycol-glycerolipids which differ mainly in the nature of their head groups.

Lipids are formed from fatty acids and alcohol by condensation. Fatty acids consist of an acidic group or carboxyl group (–COOH) with the general formula RCOOH where R may be a hydrogen molecule or a group like $-CH_3$, $-C_2H_5$ and so on. Mostly the fatty acids occur in the form of long chains consisting of carbon and hydrogen atoms, thus constituting a hydrocarbon tail where the carbon atoms are in even numbers between 14 and 22 as in Figure 2.31. The hydrocarbon tails are hydrophobic in nature and they determine the properties of lipids. The fatty acids without double bonds are called **saturated fatty acids** (palmitic acid). Sometimes, the fatty acids may contain one or more double bonds (C=C) as in Figure 2.32 and are called **unsaturated fatty acids** (oleic acid).

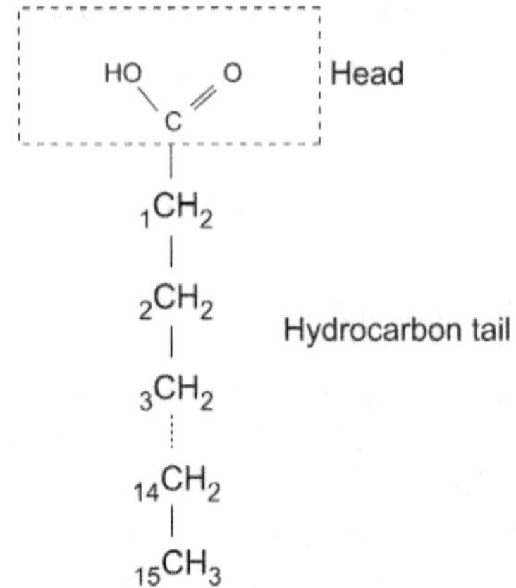

Figure 2.31. A saturated fatty acid chain (palmitic acid).

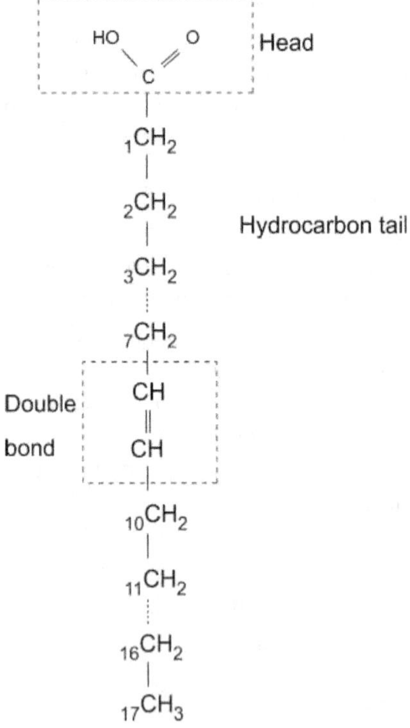

Figure 2.32. An unsaturated fatty acid (oleic acid).

Most of the lipids occur as **triglycerides**, which are formed from glycerol containing three hydroxyl groups and a fatty acid through condensation as shown in the Figure 2.33. Triglycerides are nonpolar and hydrophobic in nature, as they possess a charge within them. In other words, they are insoluble in water.

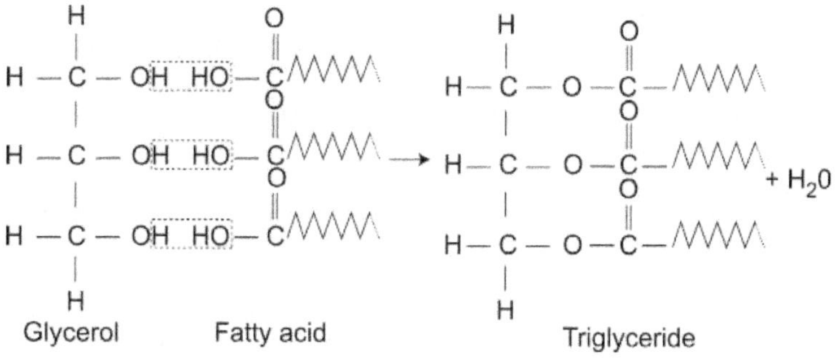

Figure 2.33. Formation of triglyceride.

2.10.1 Phospholipids

Lipids containing phosphate groups are called phospholipids. These are the building blocks of biomembranes and are assembled through non-covalent bonds. A phospholipid is formed by the combination of a phosphoric acid molecule in one of the three –OH groups of glycerol and the remaining two –OH groups with fatty acids as in Figure 2.34. Thus a phospholipid molecule has a phosphate head which is hydrophilic and water-soluble as it carries an electrical charge and hydrophobic fatty acid tail which is water-insoluble. Therefore when the phospholipid molecules are arranged in a single layer, the nonpolar hydrophobic tails project out of water and the polar hydrophilic heads arrange at the surface of water as shown in the Figure 2.35.

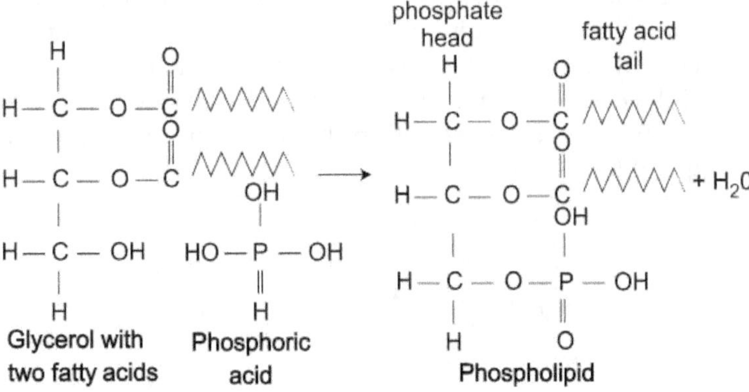

Figure 2.34. Formation of a phospholipid molecule.

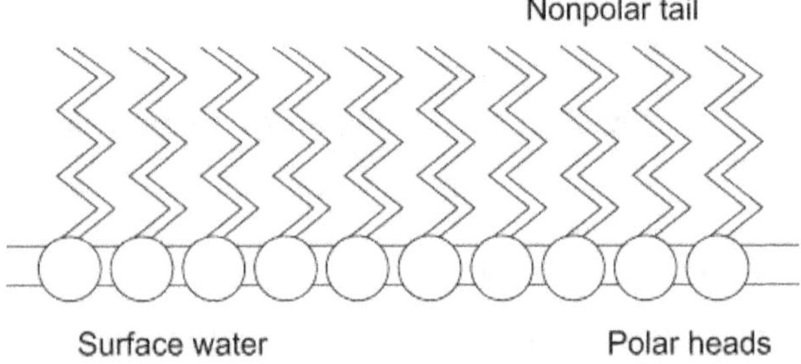

Figure 2.35. Phospholipid molecules on the surface of water.

On mechanical dispersion of phospholipids in aqueous solution, they aggregate into either spherical **micelles** or **liposomes** or **phospholipid bilayers** as shown in the Figure 2.36. The phospholipid bilayers are formed spontaneously in cells in which the fatty acid chains prevent the contact with water by linking tightly together at the centre of the bilayer, thus forming a hydrophobic core. The close linkage of the nonpolar tails is stabilized by hydrophobic and van der Waal's interactions. As the phospholipid bilayers possess hydrophobic cores, they are impermeable to salts, sugars and other small hydrophilic molecules.

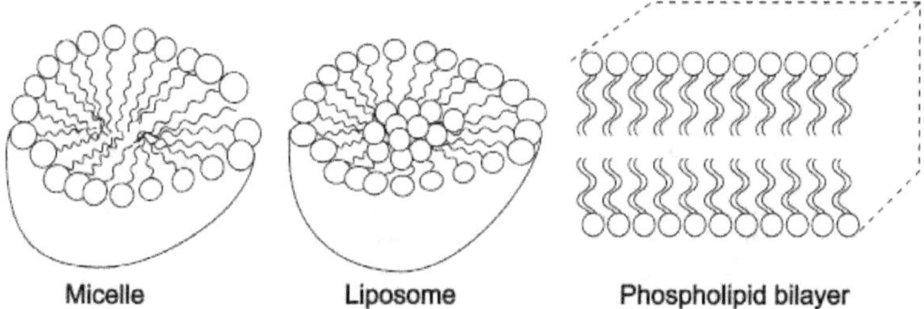

Figure 2.36. Diagrammatic representation of three forms of phospholipid aggregates.

2.10.2 Lipoproteins

Lipoproteins are soluble aggregates of lipids and proteins (lipid–protein complexes) involved in lipid transport in biological systems. Based on the density, the lipoproteins are classified into different groups namely very low-density lipoproteins (VLDL), intermediate-density lipoproteins (IDL), low-density lipoproteins (LDL) and high-density lipoproteins (HDL). However, all lipoproteins possess common structure as in Figure 2.37. Each group of lipoprotein has characteristic apoprotein with distinct lipid composition, thus forming apolipoproteins. The lipoproteins are spherical in shape with an inner hydrophobic core containing lipid and polar amino acid residues. The outside of the molecules contains hydrophilic

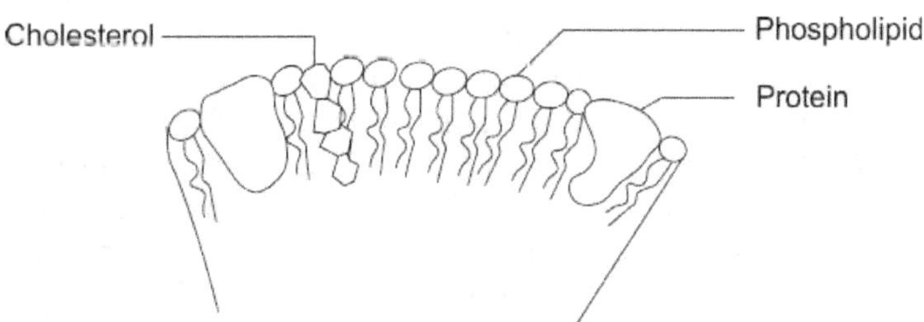

Figure 2.37. Typical structure of a plasma lipoprotein (Schematic).

protein structures and polar head groups. As an example, human lipoproteins include nine major apolipoproteins, the characteristics of which are shown in Table 2.2.

Table 2.2. Apoproteins in plasma lipoprotein of man.

Apoproteins	Molecular weight	Lipid composition
A –I	28300	HDL
A –II	17400	HDL
B –48	241000	chylomicrons
B – 100	513000	LDL
C – I	7000	Chylomicrons
C – II	10000	VLDL
C – III	9300	Chylomicrons, VLDL and HDL
D	35000	HDL
E	33000	VLDL, LDL and HDL

NUCLEIC ACIDS

2.11 ORGANISATION OF NUCLEIC ACIDS

In living cells, nucleic acids are the most fundamental and the main information- carrying macromolecules with a remarkable potential for self-duplication. Nucleic acids are of two types namely deoxyribonucleic acid (DNA) and ribonucleic acid (RNA). They are polymers formed by monomers called **nucleotides** each of which consists of a phosphate group linked by a phosphoester bond to a pentose sugar which is in turn linked with a base as shown in the Figure 2.38.

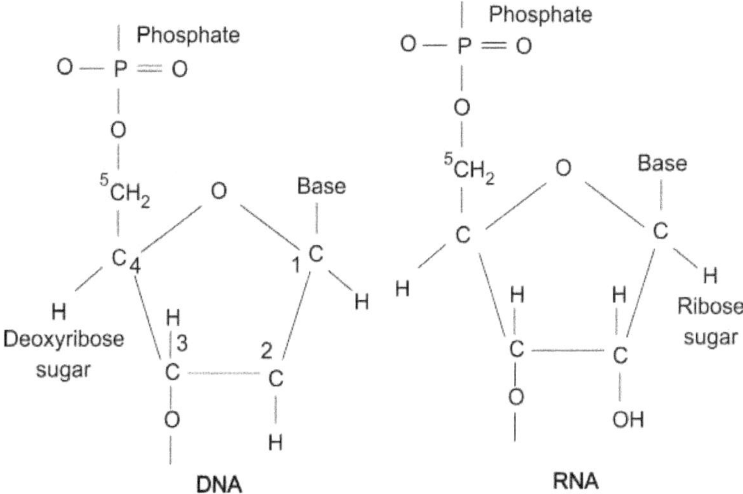

Figure 2.38. Repeating units of nucleic acids.

In each type of nucleic acid, the monomer consists of a 5-carbon sugar namely deoxyribose in DNA and ribose in RNA. The difference in the sugar molecule is due to the hydroxyl group in RNA which is replaced by hydrogen in DNA. The successive monomers are connected by phosphate residues through phosphodiester bonds between the two sugar residues as in the Figure 2.39. The phosphodiester-linked sugar residues form the back bone of the nucleic acid molecules.

Each monomer of the nucleic acid chain carries a heterocyclic base which is of two types namely **purines** and **pyrimidines**. In DNA, the purines are adenine and guanine and the pyrimidines are cytosine and thymine. The same bases are also found in RNA molecule except that uracil replaces thymine. Thus, the nucleic acids are polymers made up of four kinds of monomers. The bond between the carbon atoms of the sugar and the nitrogenous base is called **glycosidic bond** as shown in the Figure 2.40 and the monomers are called **nucleotides**. As the nucleic acids are polymers of nucleotides, they are also called as **polynucleotides.** Moreover, the bases are **tautomeric forms** in which the location of hydrogen atoms and double bonds differ and they are capable of interconversion.

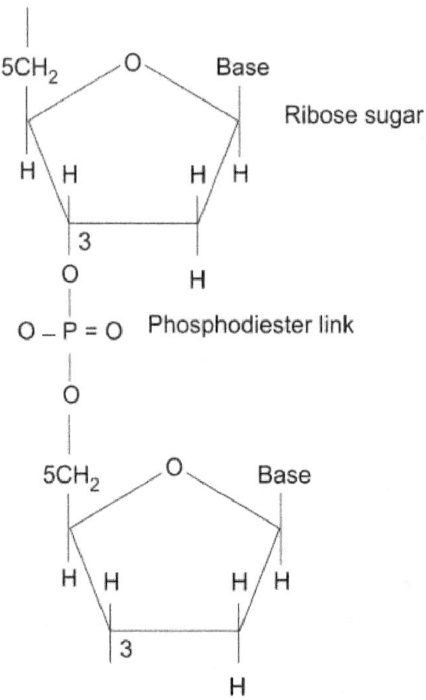

Figure 2.39. Phosphodiester bond between two sugar residues.

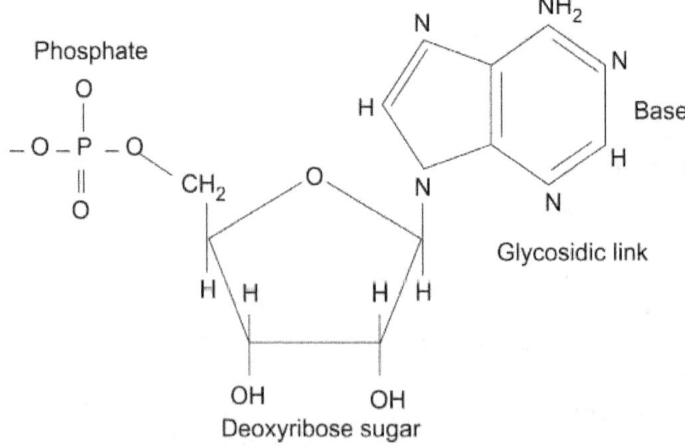

Figure 2.40. Adenine nucleotide showing glycosidic bond.

2.11.1 Primary Structure of DNA

The definite sequence of a DNA polynucleotide denotes its primary structure in which the phosphodiester bonds are present between 3′ hydroxyls and 5′ phosphates so that the molecule has a phosphate group at its 5′ end and a hydroxyl group at its 3′ end as in the Figure 2.41.

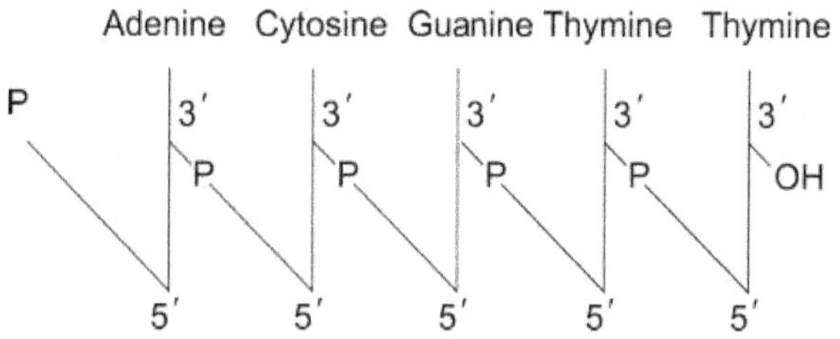

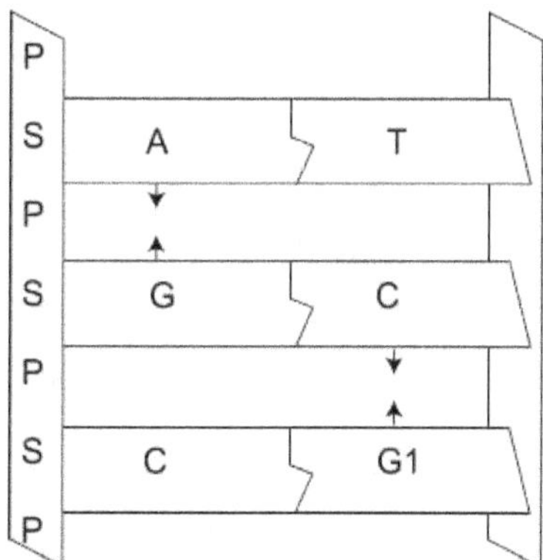

Figure 2.41. Sequence of a DNA polynucleotide.

Primary structure of DNA stores the genetic information. That is a gene is nothing but a particular DNA sequence in which the information is encoded in a four letter language (bases).

2.11.2 Secondary Structure of DNA

The secondary structure of DNA represents the formation of double strands and further coiling of DNA chain due to the

Figure 2.42. Base pairing in the two strands of DNA.

formation of hydrogen bonds between the bases of the two chains. During double strand formation, the base adenine of one chain always pairs with the thymine of the second chain and the guanine of the first chain pairs with cytosine of the second chain. The pairing of bases occurs at the same place so that the bases remain perpendicular to the sugar phosphate backbone. Therefore, the base pairs resemble the rungs of a ladder as shown in the Figure 2.42.

This double-stranded DNA molecule gets coiled and produces a right-handed double helix due to the hydrophobic

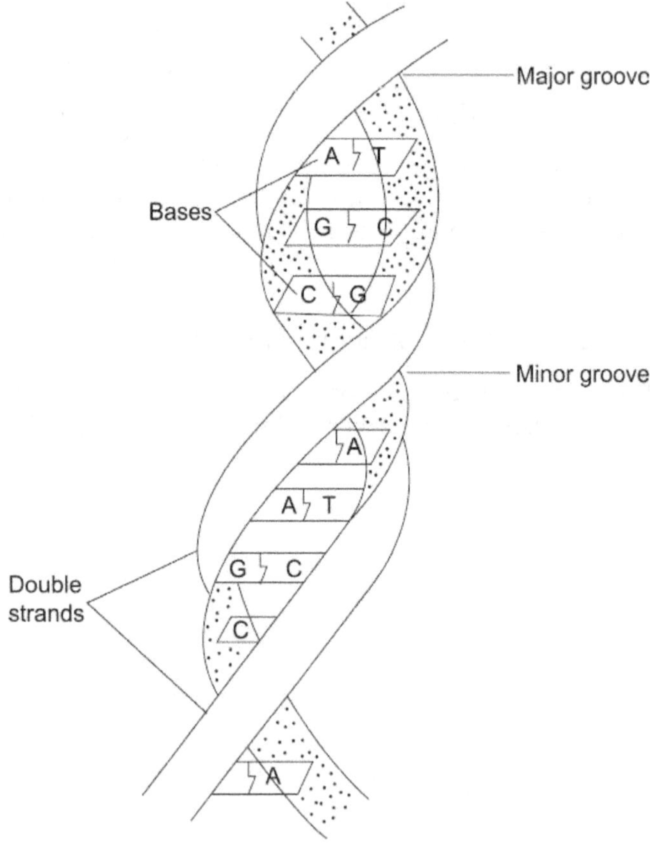

Figure 2.43. Twisting of two strands of DNA.

forces developed among the base pairs. During coiling, one chain runs with its phosphodiester bond towards the 3′ 5′ direction while the other chain runs antiparallel towards the 5′3′ direction. The attraction between the base pairs causes twisting of the two strands in opposite directions as shown in the Figure 2.43.

At the centre, all the base pairs will be in the same line so that a double helical structure results. The two strands are stabilized by hydrogen bonds between the bases. Adenine is linked to thymine by two hydrogen bonds while guanine is linked to cytosine by three hydrogen bonds as shown in the Figure 2.44. In an aqueous environment, the hydrophilic phosphate–deoxyribose backbones are on the outside in the helix and the base pairs are piled on one another with their planes perpendicular to the helical axis. The piling of bases produces strong van der Waal's interaction between them. Though the bases are inside the helix, the DNA strands exhibit two deep spiral grooves namely **major** and **minor grooves** as shown in Figure 2.43. The major grooves give more access to the bases whereas the minor grooves to the sugar backbones. As adenine always pairs with thymine and guanine with cytosine in a DNA molecule, the two strands are complementary.

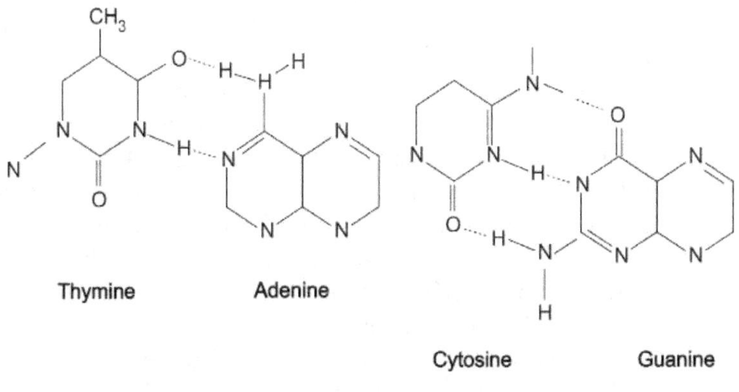

Figure 2.44. Hydrogen bonds between base pairs.

X-ray diffraction studies indicate that the DNA molecule can exist in two other forms namely B form and A form. When DNA is extracted under high humidity, then it exhibits B form and under low humidity A form is obtained. The A and B forms of DNA molecules are right-handed helices but they have different structures. In the B helix, the bases lie very close to the helical axis which passes between the hydrogen bonds. In the A helix, the bases lie outside and are strongly tilted with respect to the axis. In the B helix, the major and minor grooves are distinguishable whereas the two grooves are more or less equal in the A helix. In an aqueous environment, B-DNA is favoured but not A-DNA. This is because the B-DNA accommodates water molecules in the minor grooves and the hydrogen bonds between water molecules and DNA give stability to the B form.

Conformational transitions in secondary structure of DNA   In addition to the three secondary structures of DNA namely random coil, B form and A form, there are many conformational transitions in the DNA molecules. By X-ray diffraction method, a double-stranded helix having G–C base pairing with left hand structure has been discovered. This molecule has two most stable orientations of the bases with respect to the deoxyribose sugar rings namely 'syn' and 'anti'. While all the bases are in the 'anti' orientation in both A and B form of DNA, the pyrimidines in the left hand structured helix are always 'anti' and the purines are always 'syn'. Moreover, the repeating unit is not one but two base pairs, thus giving a zig zag pattern to the phosphate molecules. Therefore, this type of DNA molecule is given the name **Z-DNA.** Some single-stranded DNA molecules exhibit self-complementary base pairing in which the chain folds back on itself to form hairpin configuration. Some DNA sequences may form double hairpins and are called **cruciform structures**. Many polynucleoside strands of DNA can form triple helices which possess a **Hoogsteen-type** base pair in addition to the normal Watson–Crick base pairing as shown in Figure 2.45. This unusual structure incorporating a triple helix is called **H-DNA** which requires one all-pyrimidine strand and one all-purine strand.

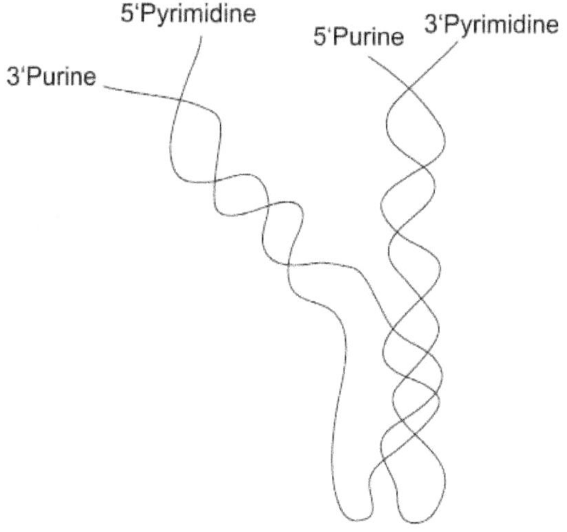

Figure 2.45. Schematic representation of a triple-stranded DNA helix.

2.11.3 Tertiary Structure of DNA

Many naturally occurring DNA molecules are circular in which free 5′ or 3′ ends are absent. They may be single-stranded or double-stranded. In the circular DNA molecule, the ends are covalently joined to form a closed circular molecule. Most of the circular DNA are supercoiled in which the molecule has extra twists in the helical axis itself. The three-dimensional structure such as supercoiling involving a higher order of folding of regular secondary structure is called tertiary structure. When the two-stranded helix is right-handed and the twisting of the strand is also right-handed, the DNA is said to be positively supercoiled and if the twisting is left-handed, then the molecule is said to be negatively supercoiled (Figure 2.46). Supercoiling of DNA molecules is an endergonic process and all the molecules isolated from higher organisms are negatively supercoiled. Supercoiling is also evident in the DNA coils which surround the basic proteins namely the histones.

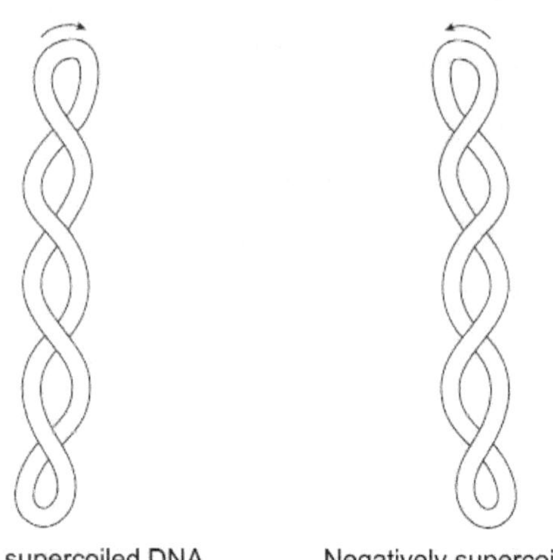

Figure 2.46. Supercoiling of DNA.

2.11.4 Structure of RNA

Though RNA molecules are generally single-stranded, they can also form hairpin structures as well as well defined tertiary

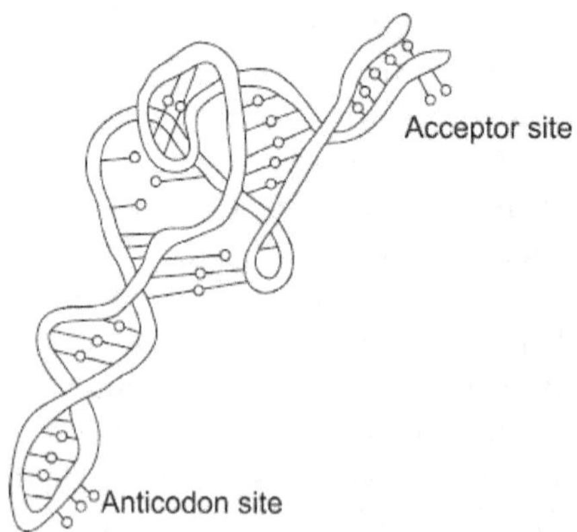

Figure 2.47. Tertiary structure of t-RNA.

structures. The secondary and tertiary structures are exhibited only by tRNA molecules. The secondary structure of these molecules is due to the folding of the chain on itself. The tertiary structure is due to the formation of intramolecular hydrogen bonding and **hydrophobic interactions** so that the helices show more complex folding as in Figure 2.47.

2.11.5 Sequencing of Nucleic Acids

In recent years, there are remarkable advancements in determining the nucleic acid sequences. Maxam–Gillbert sequencing method is the pioneering work in this field. Fred Sanger has introduced a sequencing method which is now universally used. In addition, sequencing of longer fragments of nucleic acids can be done more rapidly by automation.

Maxam–Gilbert method   This method involves the cleavage of a 5′ end-labelled DNA molecule into a series of fragments with varying lengths. Then the fragments are analysed by gel electrophoresis. In this method, the enzymes are used for selective cleavage of nucleic acids at A, T, G, or C residues so that oligonucleotide fragments are obtained at specific base positions, the sequence of which are analysed by gel electrophoresis.

Fred Sanger method   This technique is an alternate method to the Maxam–Gilbert method. Here the fragment to be sequenced is cloned into a vector namely bacteriophage M_{13}. The double-stranded replicative form of bacteriophage is isolated and cloned with the target sequence with the help of restriction enzymes. Then the cloned double-stranded DNA is introduced into a bacterium which now produces viral particles containing single-stranded DNA. These single-stranded circular DNA become the templates for four DNA polymerase-catalysed reactions in the presence of deoxyribonucleoside triphosphate analogues or the dideoxy analogues. These analogues terminate the chain extension in the reaction mixture. It is important that the enzyme reaction must occur in the presence of equivalent concentrations of dATP, dCTP, dGTP and dTTP along with ddATP of one-tenth concentration. When the reaction is terminated, the chain elongation stops and the fragments are released. In this way, a series of fragments with varying length are produced. The polymerization mixture is radiolabelled and the resulting fragments are subjected to gel electrophoresis and autoradiography to get four base specific sequencing charts.

Automated sequencing   It is a recently introduced method to sequence the nucleic acids by the Sanger method. In this technique, the sequencing reactions are carried out with differently coloured primers which are derived at its 5′ end with a fluorescent dye. This gives characteristic fluorescence to the fragment terminated by A, T, G or C. As a result, the sequence of the fragments can be determined without using radioisotopes. The sequence gel is read with the help of a computer and processing the data. More recent methods involve direct sequence determination in double-stranded DNA without cloning them into a vector.

Sequencing of transcription points   In prokaryotes, the first nucleotide of transcripts contains a 5′ triphosphate terminal which can be easily identified. This requires purification of the transcript by enzymes. The fungal enzymes S_1 nuclease cleavages the single-stranded DNA and RNA. The fragment containing the template for 5′ end of the transcript is cleaved asymmetrically using restriction enzyme so that the template DNA strand is labelled. This labelled fragment is then denatured and hybridized to mRNA which will result in a double-stranded DNA–RNA hybrid. When this hybrid molecule is treated with S_1 nuclease, a double-stranded labelled DNA is produced. The length of the DNA strand is the distance from the 5′ end of the transcript to the restriction site. By denaturing the hybrid, this distance can be identified by gel electrophoresis by keeping a set of Maxam–Gilbert cleavage fragments along the side.

2.11.6 Nucleoproteins

Nucleoproteins are formed by the combination of simple basic proteins namely histones or protamines with nucleic acids. These are conjugated proteins having high molecular weight. In eukaryotic chromosome, the negatively charged DNA is closely associated with positively charged histones to form chromatin. The chromosomes are the entangled mass of fibres of DNA–protein complex in which the DNA is linked to special proteins. The DNA-binding proteins are of two types namely histones which include H1, H2A, H2B, H3 and H4 groups and nonhistone proteins such as nuclear enzymes like polymerases, regulatory proteins, etc. Though the prokaryotic cells have DNA-associated proteins, their proteins are entirely different from the histones. Proteins and DNA are macro ions possessing net charges. That is why positively charged proteins strongly interact with DNA molecules which are negatively charged (Figure 2.48).

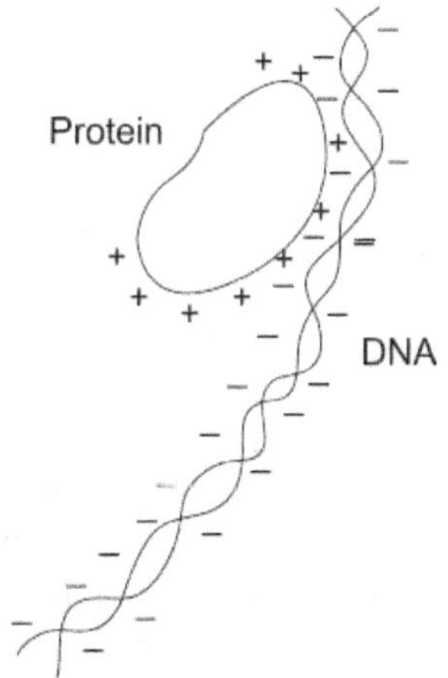

Figure 2.48. Interaction of protein with DNA.

The combination of DNA and histones forms bead-like structures called **nucleosomes**. In a nucleosome, the histones form a protein core around which the DNA chain is wound like thread as in the Figure 2.49. The core is an octamer containing two copies each of histones H2A, H2B, H3 and H4. The histone core contains flexible amino terminals containing 11 to 37 residues and these terminals are called histone tails which give a string conformation to the chromatin. The histone tails of one nucleosome may interact with the neighbouring nucleosomes or with other proteins associated with chromatin.

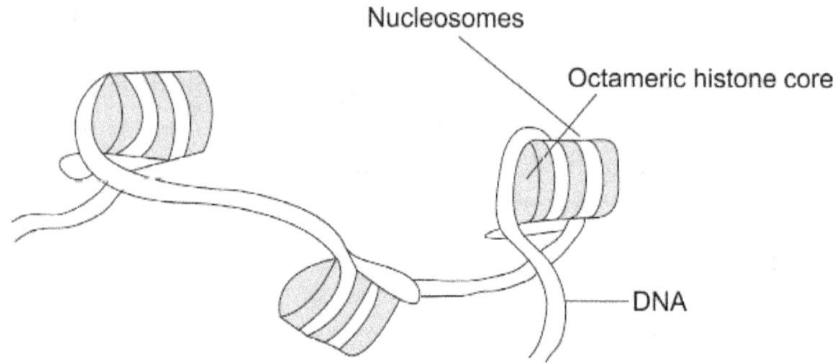

Figure 2.49. Diagrammatic structure of nucleosomes.

2.12 MACROMOLECULES OF IMMUNE SYSTEM

The immune system includes a number of specialized cells which evoke an immune response against the foreign substances or immunogens. The immunogens are otherwise called as **antigens** which stimulate the immune cells namely T cells, B cells and macrophages. The immune response results in the generation of **antibodies** or **immunoglobulins**, which bind with the triggering agents namely antigens and eliminate them. The immune system is also capable of differentiating 'self' and 'non-self' through **human leukocyte antigens** (HLA) found on the surface of the cells, which discriminate the foreign antigens.

2.12.1 Antigens

Except **haptens**, all antigens are multivalent because they possess a large number of class determinants or **epitopes** as reactive sites. The antigens are mostly protein in nature, the immunogenicity of which not only depends on the amino acid sequence but also on their three dimensional conformation. In these antigens, the reactive sites consist of six or seven amino acid residues in which the charged or aromatic side chains dominate. The determinants formed by amino acids are either close to each other or are brought close together by folding of the chain as in the Figure 2.50. The primary structure of protein with linearly arranged amino acid residues also contain epitopes. The folding in the tertiary structure of proteins results in the formation of loops lying very close together so that they possess potential immunogenicity. The antigenic determinants of polysaccharides contain six or seven monosaccharide units in which the antigenicity is not produced by the repeats but by the orientation of the repeats due to α or β bonds.

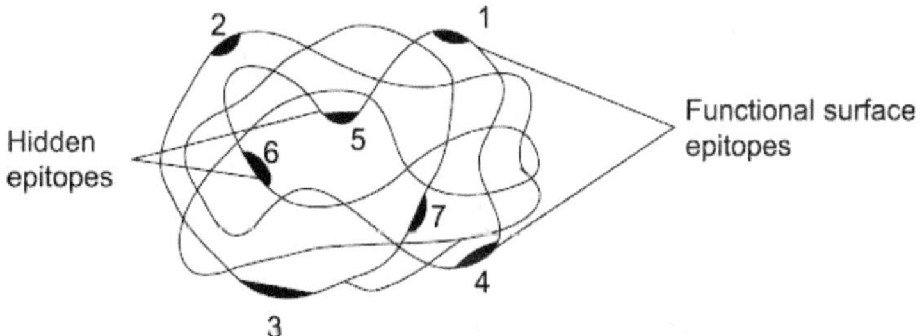

Figure 2.50. Class determinants or epitopes in an antigen molecule.

Histocompatibility antigens Many histocompatibility antigens are formed by glycoprotein chains in which the incorporation of polypeptide chains resemble immunoglobulin domains. There are two major classes of histocompatibility antigens in mammals namely MHC class I antigens and MHC class II antigens. Each MHC class I antigen consists of a heavy chain with three domains (α_1, α_2, α_3) lying outside the cell membrane and a light chain without loop (β-2 microglobulin) occurring within the cell. Each MHC class II antigen consists of two polypeptide chains namely an α chain and a β chain each of which possesses two domains (α_1 and α_2, β_1 and β_2). These domains exhibit disulphide linkage and are extracellular whereas the main chains are embedded in the cell membrane with their hydrophobic ends facing outside.

2.12.2 Antibodies

Antibodies are the major macromolecules of the immune system and are protein in nature having structural similarity with globular proteins. Therefore, they are given the name immunoglobulins. The typical antibody molecule is IgG which is formed by four peptide units namely two identical heavy chains and two light chains. The molecule is a glycoprotein as the heavy chains are associated with carbohydrates. The light chains contain about 220 amino acid residues while the heavy chains contain about 440 amino acid residues. The four chains are linked by disulphide bonds in the form of intra-chain and inter-chain disulphide linkages at the region of domains (Figure 2.51.) The molecule exhibits a tertiary structure in which the variable and constant regions are formed by β-pleated sheets. Both intra-chain disulphide bonds and hydrophobic interactions in the amino acid residues stabilize the structure of the molecule.

Isotypes of immunoglobulins   Based on the nature of heavy chain constant regions, immunoglobulins are classified into IgG, IgA, IgM, IgD and IgE, which thus form isotypes. They differ from each other in their amino acid sequences at their constant regions so that the constant region of IgG is denoted by γ (gamma), IgA by α (alpha), IgM by μ (mu), IgD by δ (delta) and IgE by ε (epsilon). IgA molecules are

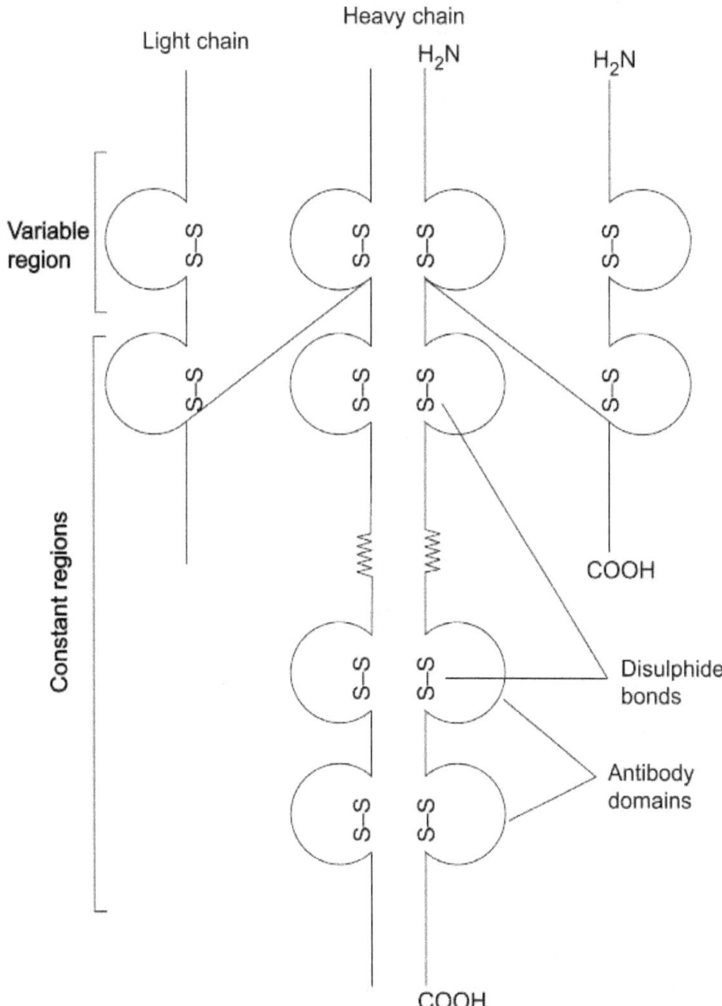

Figure 2.51. Antibody domains with intra- and inter-chain disulphide linkages.

characterized by having higher carbohydrate content and more disulphide bonds in heavy chains when compared to IgG molecules, but the bonds occur in different positions. IgA molecules also occur as dimers. IgM molecules contain higher carbohydrate content in their heavy chains than in IgG and IgA molecules and mostly occur as pentamers. IgD and IgE molecules are not capable of polymerization and they occur only as monomers.

REVIEW QUESTIONS

1. Define dipole, polarity and dipole moment.

2. What are polar and apolar molecules?

3. Explain various dipole interactions.

4. How do macromolecules maintain stability?

5. Write an essay on different types of bonds in biological molecules.

6. Explain covalent and non-covalent bonds.

7. Write notes on ionic bonds, hydrogen bonds, van der Waal's interactions and hydrophobic effects.

8. Give an account of lipid and protein composition of biomembranes.

9. Describe the membrane proteins on the basis of membrane protein interactions.

10. What is membrane permeability? Explain the process of transport of substances across the membrane.

11. Give an account of extracellular matrix.

12. What are collagens and multiadhesive matrix proteins?

13. Explain primary, secondary, tertiary and quaternary structure of proteins.

14. Describe the structure of insulin, keratin, fibroin, myoglobin and haemoglobin.

15. Write an account on cellular aspects of protein folding.

16. Explain the mechanism of protein–protein and protein–ligand interactions.

17. How could protein kinase be activated through ligand binding?

18. What is protein sequencing? Enumerate the methods which are used in sequencing the proteins.

19. Describe different types of bonds involved in protein denaturation.

20. Give an account of protein denaturation.

21. What are the characteristics of denatured proteins?

22. Explain the denaturing agents of proteins with examples.

23. Describe various conformations of monosaccharides.

24. Explain various bonds involved in the formation of disaccharides and polysaccharides.

25. Write a note on glycoproteins and their pattern of linkage.

26. Explain a fatty acid chain.

27. Describe the bonds involved in the formation of a phospholipid molecule.

28. Explain the typical structure of a plasma lipoprotein.

29. Write an account on primary, secondary and tertiary structure of DNA.

30. Explain the glycosidic bonds between bases and sugar, and hydrogen bonds between base pairs.

31. What are the conformational transitions which occur in the secondary structure of DNA?

32. Describe the tertiary structure of t-RNA.

33. Explain the methods employed in sequencing nucleic acids.

34. Write notes on nucleoproteins and nucleosomes.

35. Write an essay on macromolecules of immune system.

3

PRINCIPLES OF KINETICS OF MOLECULES

In living systems, permeability of cell membrane is important to maintain satisfactory intracellular physiological conditions. The plasma membrane acts as a permeability barrier and controls the movement of water and solutes in and out of the cell so that a difference is maintained between intercellular and extracellular fluids. Based on the nature of permeability, plasma membrane can be distinguished into two types namely semipermeable membrane and selectively or differentially permeable membrane. A semipermeable membrane allows the movement of a solvent like water but not a solute like colloids to pass through it. But a selectively permeable membrane not only permits water but also solutes like salts and sugars through it in varying degrees. The movement of molecules across the cell membrane occurs at all times in cells and the principles of kinetics of the molecules are dealt with in this chapter.

3.1 DIFFUSION

It is the phenomenon of passive transport or wandering of solute molecules through membranes resulting in the same concentrations of diffusing substances on both sides of the membrane. In other words, the movement of ions and molecules across the membrane from their higher concentration to lower concentration along the concentration gradient or electrochemical gradient without requiring energy is called **diffusion**. Therefore, this phenomenon is otherwise called as **passive transport**. In liquids, diffusion occurs slowly whereas in gases, it is rapid.

3.1.1 Factors Affecting Diffusion

The rate of diffusion is affected by the following factors:

1. Higher diffusion gradient or differences in concentration between two compartments increases the rate of diffusion.

2. Greater surface area of the membrane through which diffusion occurs increases the rate of diffusion.

3. Diffusion occurs only at very short distances and the rate of diffusion decreases with the distance.

3.1.2 Simple Diffusion

As the plasma membrane possesses numerous pores, small and hydrophilic substances can easily pass through these aqueous channels along the concentration and electrochemical gradient. Addition of a homogenous sugar solution to water will cause a net movement of sugar molecules depending on the concentration gradient. Let us consider that a membrane separates two solutions of a solute x with concentrations x_1 and x_2 in two compartments. Suppose the initial concentration is higher in compartment A, then there is a net transport of substances from compartment A to B until equal concentrations are obtained in both compartments (Figure 3.1). Thus a substance can diffuse spontaneously without requiring energy through a membrane from the region of higher concentration to the region of lower concentration. That is, diffusion of substances occurs from a region of high chemical potency to a region where the chemical potency is low until equilibrium is attained.

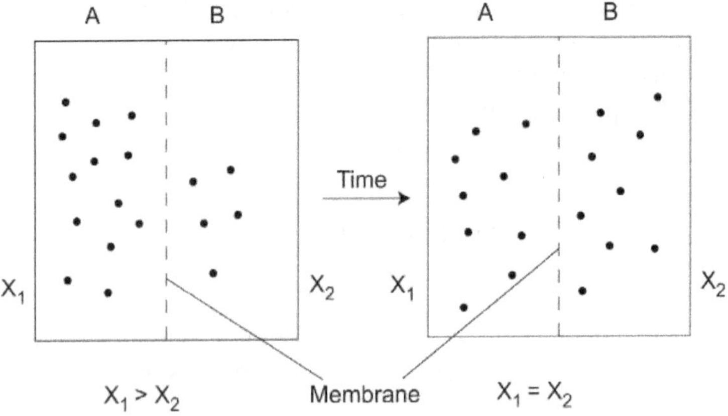

Figure 3.1. Diffusion and equilibrium of ions across a membrane.

The net rate of transport of substances by passive diffusion is directly proportional to the concentration gradient across the membrane as in the formula given below:

$$J = \frac{KD_1(C_2 - C_1)}{l}$$

where,

J = net rate of transport

K = Partition coefficient of the substance

D_1 = Diffusion coefficient of the substance

C_2 and C_1 = Concentration gradients and

l = Thickness of the membrane

However, it is not possible to measure exact values of K and D_1 and the thickness of the membrane. Therefore, the rate of diffusion can be explained in terms of permeability coefficient as

$$J = -P\,(C_2 - C_1)$$

The rate of diffusion of ions and molecules through a semi permeable membrane depends on the size of the particles, concentration gradient across the membrane and solubility of the molecules in lipid. Therefore, the permeability of substances through a semipermeable membrane (P) can be represented by the formula,

$$P = \frac{KD}{I}$$

where,

K = Partition coefficient

D = Diffusion coefficient and

T = Thickness of membrane

This clearly indicates that the molecules of equal size having higher lipid solubility can penetrate the membrane at faster rate.

3.1.3 Fick's Law of Diffusion

According to this law, the diffusion of substances depends on the concentration gradient, diffusion coefficient and area of diffusion and this can be represented as,

$$dQ = DA(dc/dx)dt$$

where,

dQ = mass flow (moles) at area (sq.cm)

D = Diffusion coefficient (sq.cm/sec)

dc/dx = concentration gradient (moles/cm)

dt = time taken (sec)

Thus, diffusion coefficient determines the rate of movement of a molecule in a solution and it depends on the size and shape of the molecules and the medium. For example, the diffusion coefficient of oxygen is 1.25 times higher than that of carbon dioxide. The rate of diffusion of oxygen and carbon dioxide in air is 100 times greater than in water. The diffusion coefficient of substances is closely related with their molecular weight.

3.1.4 Diffusion of Electrolytes

Ions are electrically charged particles, the movement of which depends on the nature of their charge. As one ion attracts the ions of opposite charge, the ions of opposite charge move in pairs. Therefore, it is difficult to derive the diffusion coefficient of a single ion. In an electric field, the ionic pairs dissociate and move independently towards their respective opposite poles. This ionic mobility of individual ions can be measured. In an electrolyte solution, the anions and cations move at different rates across a membrane resulting in the formation of a potential. This occurs when two solutions of the same electrolyte with same or with different concentrations are separated by a membrane. Such potentials are called **diffusion potentials** which depend on the nature of membrane and the ions. In living cells, the ionic concentration of the cytoplasm in relation to the extracellular medium and the electrical potential across the membrane are closely related. This is because the electrical potential across the membrane depends on the unequal distribution of ions on both sides of the membrane. As the ions are charged particles, their diffusion depends on the electrical gradient in addition to the concentration gradient. Thus any change in the membrane potential would alter the ionic concentration on either side of the membrane.

3.1.5 Facilitated or Accelerated Diffusion

The rate of transport of many substances by simple passive diffusion is insufficient for normal functioning of living cells. Therefore, some special mechanisms are required to increase the diffusion of substances. This is achieved by facilitated diffusion which is a carrier-mediated transport mechanism involving specific carrier molecules, mostly specific proteins to carry the substances across the membrane along the concentration gradient without energy requirement. This includes two types namely pore-facilitated diffusion and carrier-facilitated diffusion.

Pore-facilitated diffusion   In the membrane of red blood corpuscles the transmembrane protein forms an ion pore or anion channel as shown in Figure 3.2. This channel is permeable to Cl and HCO_3 ions and is highly selective. Thus it allows only the exchange of Cl and HCO_3 ions so that the transport of HCO_3 ions from the erythrocytes is balanced by the influx of Cl ions.

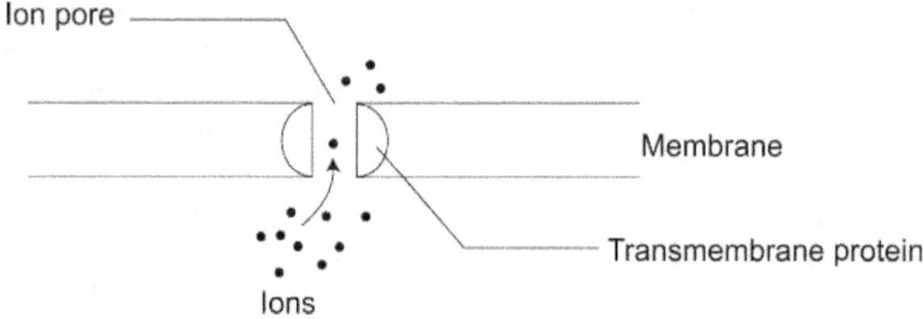

Figure 3.2. Pore-facilitated diffusion across membrane.

Carrier-facilitated diffusion   Here, a carrier molecule found in the membrane becomes linked with the substances which it carries along as shown in the Figure 3.3. The carrier molecule is capable of free movement from one surface to another surface of the membrane. Therefore, the substance when attached to the carrier molecule at one surface of the membrane gets released at the opposite side of the membrane.

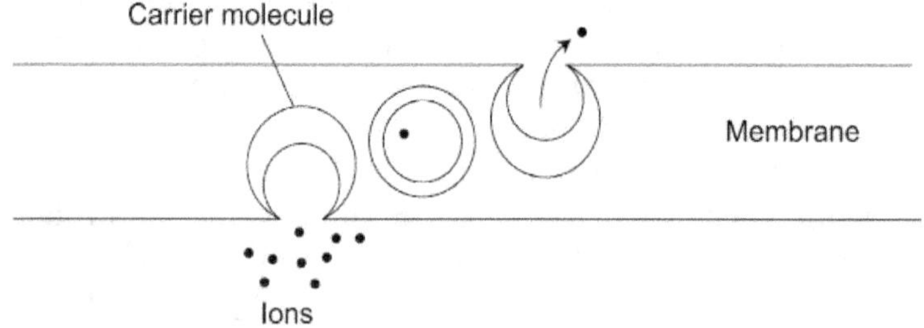

Figure 3.3. Carrier-facilitated diffusion across membrane.

3.1.6 Biological Significance of Diffusion

1. In nerve impulse conduction, the membrane potential develops by the diffusion of ions through the membrane as a result of an imbalance of negative and positive charges on each side of the membrane.

2. Diffusion plays an important role in the intestinal absorption of pentoses, minerals and water-soluble vitamins and in the renal absorption of urea.

3. Diffusion plays a vital role in the whole process of respiration.

4. Protein cations of plasma membrane are responsible for the diffusion of cations and for increasing the anions.

3.2 OSMOSIS

Osmosis is the movement of water molecules from a region of their higher concentration to a region of their lower concentration through a semipermeable membrane. It is the passage of a solvent from its higher concentration to its lower concentration through a semipermeable membrane. Strictly speaking, it is a kind of diffusion in which only water molecules move. The osmotic flow of the solvent occurs when a membrane separates two solutions of the same nature but differing in concentrations until the concentrations of the solutions become equal. In other words, there is movement of excess of solvent from a weaker solution to a stronger solution till an equilibrium is attained. Therefore, it is a passive process wherein the movement of the solvent depends on the concentration gradient of two solutions.

The tendency of water molecules to move from one region to another is referred to as **water potential** (Ψ-psi) so that water always moves from a region of higher water potential to a region of lower water potential. As water molecules possess kinetic energy, a system with greater concentration of water molecules will have higher water potential. Therefore, osmosis can also be defined as the passage of water molecules from a region of higher water potential to a region of lower water potential through a semipermeable membrane. The concentration of dissolved solute molecules reduces the concentration of water molecules in a solution, thus lowering its water potential. That is why all solutions possess lower water potential than pure water. The extent of lowering the water potential by the solutes is called **solute potential** (Ψs). The pressure required to move the water molecules from one region to another is called **pressure potential** (Ψp) which increases the water potential.

3.2.1 Osmotic Pressure

The hydrostatic pressure required to prevent the flow of solvent molecules through a semipermeable membrane is called **osmotic pressure**. It is the pressure applied to a solution to restrict the flow of solvent through a membrane. Osmotic pressure is characteristic of dilute solutions. Vant Hoff observed a striking similarity between the properties of dilute solutions and gases. According to his theory (**Vant Hoff's theory of dilute solutions**), a substance in a solution behaves exactly like a gas and the osmotic pressure of a dilute solution is equal to the pressure, which the solute would exert if it were a gas at the same temperature and occupies the same volume as the solution. Thus osmotic pressure is the pressure exerted on a solution in order to increase its vapour pressure until it becomes equal to that of the solvent.

The osmotic pressure of a solution is directly proportional to the concentration of the solute, which in turn depends on the number of molecules or ions present in the solution. If a substance has lower molecular weight, then it will have more moles per unit volume than a substance with higher molecular weight so that it will exert a greater osmotic pressure. If the osmotic pressure of two solutions is the same, then they are referred to as **iso-osmotic solutions** and they also possess the same vapour pressure. If a solution exhibits greater osmotic pressures than another one, then it is called as **hyper-osmotic solution.** On the other hand, if a solution has lesser osmotic pressure than another, then it is **hypo-osmotic solution.**

3.2.2 Laws of Osmosis

According to the first law of osmosis, at constant temperature the osmotic pressure of a dilute solution is directly proportional to the concentration of dissolved substances. This phenomenon is explained by citing the example of cane sugar in aqueous solution having different concentrations at constant temperature (15°C) as in Table 3.1.

Table 3.1. Osmotic pressure of the aqueous solution of cane sugar at different concentrations.

Concentration (C) (gm/100 cm^3)	Osmotic pressure (p) (Hg level, cm)	p/C
1.003	52.1	52.0
2.014	102	50.2
2.767	152	55.0
4.060	209	51.5
6.138	307	50.0

According to second law of osmosis, the osmotic pressure of a dilute solution is directly proportional to its absolute temperature and is explained in the following Table 3.2.

Table 3.2. Osmotic pressure of the aqueous solution of cane sugar with constant strength at different temperatures

Absolute temperature (T) (T°K)	Osmotic pressure (P) (Hg level, cm)	p/T
279.8	50.5	0.180
286.7	52.5	0.183
295.0	54.8	0.186
305.0	54.4	0.178
309.0	56.7	0.184

3.2.3 Determination of Osmotic Pressure (Osmometry)

Osmotic pressure is characteristic of dilute solutions and it can be measured by **osmometers** or by calculations. The simple instrument to measure directly the osmotic pressure of solutions is mercury osmometer.

Mercury osmometer It consists of a glass tube with a porous bottom coated with copper ferrocyanide which serves as a semipermeable membrane. The tube is attached to a mercury manometer as shown in Figure 3.4. The test sample is taken in the tube and the porous bottom is immersed into distilled water contained in a beaker. Now water enters into the tube through the copper ferrocyanide coating by osmosis. This produces an excess pressure in the porous bottom and this raises the mercury level in the manometer. The raise in the mercury level is the osmotic pressure of the sample.

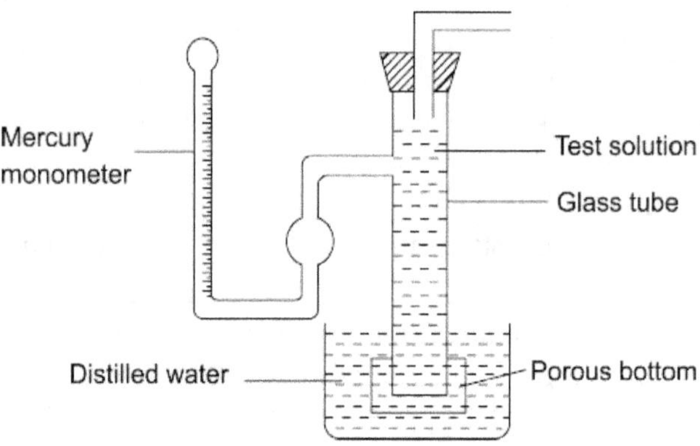

Figure 3.4. A Mercury osmometer.

Calculation of osmotic pressure from Vant Hoff's equation

Vant Hoff's equation namely pV = RT is homologous to gas equation PV = RT. According to Vant Hoff's equation,

$$\pi V = RT$$

where,

π = osmotic pressure

V = volume of the solution

R = solution constant and

T = temperature.

Suppose C gram moles of solute is present in 1 litre of solution, then 1 gram mole will be present in $1/C$ litres, and V is equal to $1/C$ ($V = 1/C$)

Therefore,

$$\pi \times 1/C = RT \text{ (or) } \pi = RTC$$

Thus if the concentration of the solute (C) is known, then the osmotic pressure of the solution can be calculated at any temperature.

3.2.4 Determination of Concentration by Depression in Freezing Point Method

Measurement of osmotic pressure is made in terms of depression of freezing point which is one of the colligative properties of solutions including osmotic pressure. The depression of freezing point is a constant and its value is $1.86°C$ (the molal freezing point depression of water). The depression in freezing point is related to the molar concentration and is shown in the equation given below.

$$\text{Molar concentration (C)} = \frac{\text{Depression in freezing point (}d\text{)}}{\text{Molar depression constant for water (}K\text{)}}$$

Then this C value is substituted in Vant Hoff's equation. The depression of freezing point of the samples can be determined using cryoscopic osmometer.

Cryoscopic osmometer   It consists of a sample vessel, the mouth of which is fitted with a measuring head having a thermister and a vibrating stirrer (Figure 3.5). The vessel is kept in a cooling chamber which may be a built-in refrigeration unit. The sample is taken in the vessel and cooled in a cooling chamber. The stirrer is used to induce freezing of the sample. When the sample freezes, its temperature rises rapidly to its freezing point. This temperature is measured by the thermister and recorded on a meter. Temperature–time curves for water and the sample are drawn on a graph sheet and the depression of freezing point (ΔT) of the sample could be derived from the graphs.

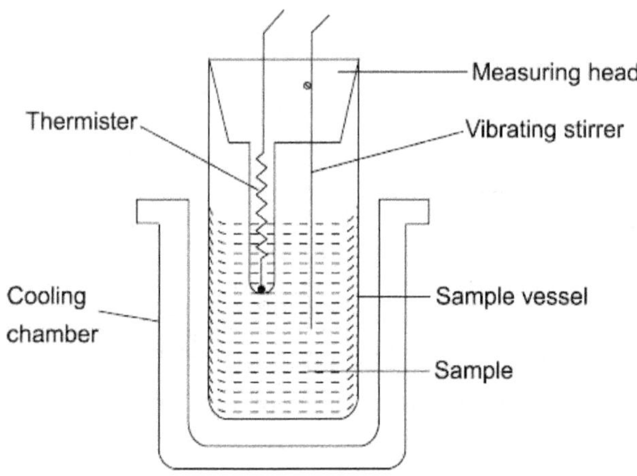

Figure 3.5. A cryoscopic osmometer.

The determination of depression of freezing point is also useful to calculate the osmotic pressure of the samples by using the following equation:

$$\frac{\Delta T}{\Delta m} = \frac{X}{22.4}$$

where,

ΔT = depression of freezing point of the sample

Δm = depression of freezing point of water

X = osmotic pressure of the sample and

22.4 = osmotic pressure of water

Therefore,

$$X = \frac{\Delta T \times 22.4}{\Delta m}$$

3.2.5 Determination of Osmotic Pressure of Electrolytes

As electrolytes dissociate to form ions, osmotic pressure of the electrolytes depends on the degree of dissociation and the number of particles present in the solution. In a particular molar concentration, they exert greater osmotic pressure than the non-electrolytes. Therefore, the osmotic pressure calculated for the electrolytes by gas law is to be incorporated by a correction factor for dissociation called **cryoscopic coefficient** (G). After the determination of freezing point depression of an electrolyte by cryoscopic osmometer, the value for G is calculated using the formula,

$$G = \frac{\Delta Tfp}{\Delta Tfb}$$

where,

Tfp = freezing point depression of electrolyte and

Tfb = freezing point depression of non-electrolyte

Then the concentration (C) is multiplied by the G value. Thus the osmotic pressure of an electrolyte $(p) = (C \times G)RT$.

3.2.6 Determination of Molecular Weight by Hydrodynamic Method

This method is useful to determine the molecular weight of high molecular weight substances like proteins. Using this method, the osmotic pressure of sample solution with different concentrations is measured and the values are substituted in the Vant Hoff's equation,

$$\pi V = RT$$

$$or$$

$$\pi = CRT$$

where C is C/M.

$$M = CRT/\pi$$

where,

M = molecular weight

C = concentration

R = gas constant

T = temperature and

π = osmotic pressure

3.2.7 Biological Significance of Osmosis

- Osmosis plays a vital role in the regulation of blood volume and in the excretion of urine.

- Osmosis works simultaneously with diffusion in the intestinal absorption.

- Osmometry is useful to evaluate the effects of drugs on renal function, to measure osmotic concentration of body fluids of animals, to determine molecular weight of macromolecules and to diagnose diseases in man.

- The osmotic pressure helps to maintain a dynamic equilibrium between the internal and external compartments of living organisms in order to maintain internal homeostasis in organs like intestine, kidney, lungs, liver, etc.

3.3 FILTRATION

Filtration is the passage of fluid through a membrane due to differential hydrostatic pressure on the two sides of the membrane. In a fluid, the molecules are in constant motion in all directions and this movement depends on the size of the molecules. In mammals, filtration plays a key role in the passage of water and electrolytes from the arterial end of the capillaries to the tissue space and in the formation of urine in kidneys. In microbiology, filtration is an effective means of removing microorganisms and other suspended matter from water.

3.3.1 Passage of Fluid through Blood Vessels into Tissue Space

In circulation, the blood supply to the tissues depends on the hydrostatic pressure. During circulation, the pressure would be highest at the central point (heart) and it will be decreasing when the blood travels to the periphery. Therefore, the pressure in the arteries is higher whereas the pressure in the peripheral vessel is always lower than the aortic pressure. The arterial capillary pressure is 22 mm Hg that is higher than the colloidal osmotic pressure of the plasma (15 mm Hg) so that the water and electrolytes travel to the tissue space from the arterial end of the capillaries. On the other hand, the colloidal osmotic pressure of plasma is 15 mm Hg which is greater than the capillary blood pressure at the venous end (7 mm Hg).

3.3.2 Formation of Urine by Filtration

In kidneys, urine is formed by ultrafiltration, which is nothing but filtration under hydrostatic pressure. In the glomeruli of kidneys, the basement membrane of the blood capillaries forms the main filtration barrier and the blood, which passes from the glomerulus, has increased concentration of plasma proteins and lesser hydrostatic pressure than that of the blood. The solute potential on either side of the filtration barrier causes water to move from less negative solute potential to more negative solute potential. As a result, the blood which passes through the glomerulus, loses water and small solutes and the plasma proteins are retained in the blood.

The formation of filtrate from the blood by the capillaries of glomeruli is due to the high capillary blood pressure (70 mm Hg) in the glomerular capillaries when compared to the blood pressure at the arterial end of the capillaries (30 mm Hg). The net balance of the forces acting on the glomerular membrane is called **effective filtration pressure,** which can be represented in the following formula:

Effective filtration = Glomerular pressure
pressure   – (Colloid osmotic pressure
                              + Bowman's capsular pressure)

The **filtration fraction** (FF) denotes the quantity of plasma, which is passing through the kidney for filtration in the glomeruli of nephrons.

$$\text{Filtration fraction} = \frac{C\text{ln}}{C\text{PAH}} = \frac{GFR}{RPF}$$

where,

CIn = Innulin clearance

CPAH = Para-aminohippuric acid clearance

GFR = Glomerular filtrate rate and

RPF = Renal plasma flow

3.3.3 Role of Filtration in Microbiology

Many waters require some type of filtration because of turbidity and colour as well as the presence of a large amount organic matter. Filtration is also used extensively to sterilize a liquid or gas. The solutions, which are easily destroyed by heat, can be sterilized by filtration.

3.4 DIALYSIS

It is the separation of particles by partially permeable or semipermeable membrane and is useful to separate lower molecular weight components from macromolecules. In other words, the process of retention of colloidal particles by a semipermeable membrane and movement of dispersion medium is called **dialysis**. The semipermeable membrane has free access to small molecules and will not allow proteins and other macromolecules. Dialysis is also used to remove contaminants of small molecules or to change the buffer condition in a system.

When a bag made up of semipermeable membranes containing a solution of protein is immersed in a buffer solution, smaller molecules with low molecular weight diffuse through the pores of the membrane. But the protein molecules are retained in side the bag (Figure 3.6). When a cellophane bag containing a dye solution of small molecules is dipped in water, then water enters into the bag and both water and dye molecules pass out from the bag. In this process, the entry of water into the bag is quicker so that a difference in osmotic pressure is developed inside and outside the bag. When a bag containing a dye solution of large molecules is immersed in sodium chloride solution, water passes out of the bag and sodium chloride ions enter into the bag simultaneously until the osmotic pressure of the salt solution equals the osmotic pressure of the salts inside. Thus removal of fluid during dialysis is due to the hydrostatic and osmotic pressure gradients across the membrane. This is the principle involved in artificial dialysis to remove the fluids.

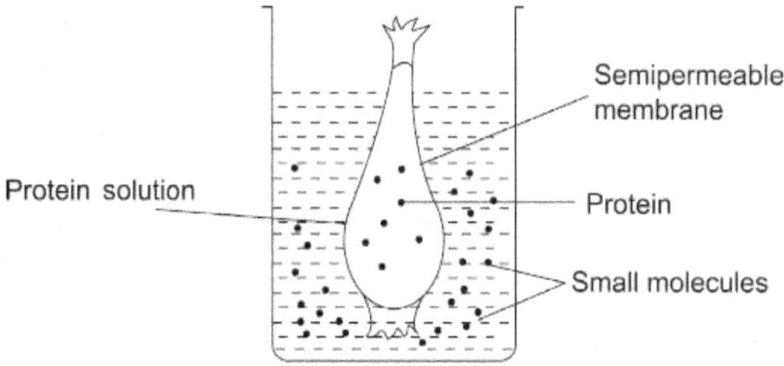

Figure 3.6. Separation of protein molecules by dialysis.

3.4.1 Principle of Dialysis in Artificial Kidney

The artificial kidney is a dialysis unit having a dialysing fluid (dialysate), which is an electrolyte solution of suitable composition. Dialysis occurs between the blood and dialysate across a perforated artificial membrane as shown in Figure 3.7.

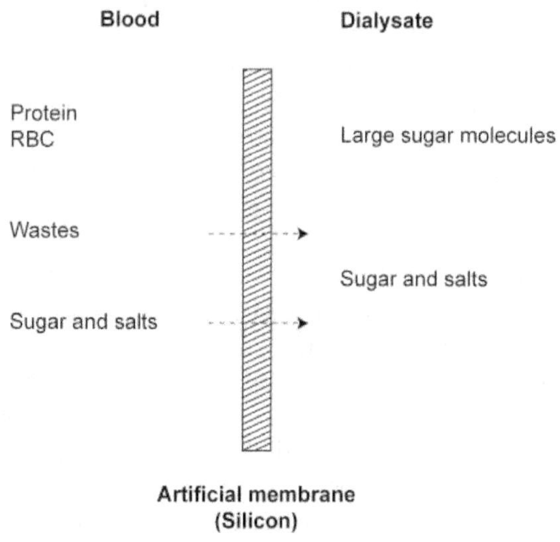

Figure 3.7. Principle of dialysis in artificial kidney.

The waste materials from the blood pass into the dialyzing fluid which is free from the waste molecules due to the existence of a concentration gradient. As a result, the waste molecules are immediately washed into the dialysate resulting in the clearance of blood. The removal of wastes during dialysis is proportional to the concentration gradient across the membrane. Thus the concentration of waste molecules in the dialysing fluid should be zero in order to have the maximum gradient. The volume of the body fluid cannot be controlled by dialysis and so a positive pressure is applied to the blood compartment or a negative pressure to the dialysate compartment. As a result, the fluid moves from the compartment of blood to the dialysing fluid. The rate of fluid removal is due to the transmembrane pressure gradient, which can be calculated by the formula

$$\text{Transmembrane pressure gradient} = \tfrac{1}{2}\,[PBi + PBo] - \tfrac{1}{2}\,[PDi + PDo]$$

where,

PBi and PBo = Blood inlet and outlet pressures and

PDi and PDo = Dialysate inlet and outlet pressure

A dialysing machine is used to produce warm dialysate, which is circulated through a dialyser assembly. It controls the cycling of blood from the patient to the dialyser as it contains monitor systems. Therefore, it is useful to calculate the quantity of fluid removed during dialysis.

3.4.2 Kinds of Dialysis

There are two kinds of dialysis namely **haemodialysis** in which an artificial membrane is used in the instrument and **peritoneal dialysis** in which a natural membrane like peritoneum in the body is used.

Haemodialysis   Here the blood is gently circulated between the artery and vein through dialysis tubing. The tube is an artificial semipermeable membrane through which ions, very small molecules and water diffuse whereas blood cells, platelets and protein molecules cannot pass. The tube is bathed in a dialysis solution so that exchange of substances occurs between the blood and the dialysing solution until equilibrium is attained. As a result, the unwanted substances such as urea and excess of sodium and potassium ions, etc. are removed and useful substance like proteins is retained.

Peritonial dialysis   Here the peritoneal membrane lining the abdominal cavity acts as a semipermeable membrane. The dialysing fluid is administered through a thin plastic tube into the abdominal cavity. As a result, exchange of substances occurs between the dialysing fluid and tissue fluid. During this process the patient can live relatively a normal life and hence the name continuous ambulatory peritoneal dialysis (CAPD).

3.5 SURFACE TENSION

In nature, matter exists in three different states namely solid, liquid and gas. If matter occurs in more than one state in a system, then each state is called a **phase** and the region of contact between the two phases is called an **interface**. The interface formed between a solid and a gas or between a liquid and a gas is called **surface**. In the interior of a phase, the molecules are always attracted to each other as same molecules surround them. The attractive force of the molecules is uniform in all directions and so the molecules move freely in all directions. But the molecules are partially surrounded by other molecules at the surface and so they are inwardly attracted, with the resultant minimization of the number of molecules at the surface. The force with which the surface molecules of a solution are held together is called 'surface tension'. It is defined as the force in dynes acting at right angles to any imaginary line of 1cm length on the surface. The surface molecules of a solution are attracted only downwards and sideways but not upwards. Thus they are held together on the surface of the liquid without free movement.

3.5.1 Kinetic Theory of Surface Tension

The surface tension occurring at the surface of the separation point of two immiscible phases (liquid–liquid or liquid–solid) is called **interfacial tension**. The kinetic theory of surface tension can be explained by taking an example of 5 molecules A, B, C, D and E which occur near the surface of a liquid with their spheres of influence (force) as shown in Figure 3.8. In molecule A, the sphere of influence is fully within the liquid, so it experiences no attraction in any direction. In molecule B which lies near the surface, a part of its force lies outside the surface. Therefore, its upward pull will be lesser than its downward pull and so will be subjected to a downward pull. On the other hand, the upward pull will be higher than downward pull in molecule D. As the molecule E lies away from the liquid, it is free as a vapour. Thus, the force is maximum on the molecules at the surface and the molecules of a liquid are subjected to downward pull. In a liquid, the molecules move from the interior towards the surface but are prevented from escaping into the space above the liquid due to a strong downward force.

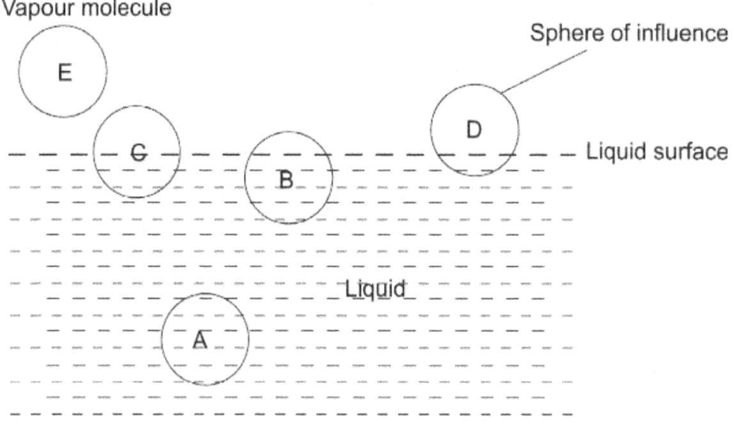

Figure 3.8. Kinetic theory of surface tension.

3.5.2 Factors Affecting Surface Tension

1. The surface tension of a liquid decreases with the increasing temperature. When the temperature of a liquid is increased, the kinetic energy of the molecule is increased, thus reducing the force between the molecules. Therefore, the inward pull on the surface molecules is lowered thereby reducing the surface tension of the liquid.

2. Many substances like proteins, organic acids, ethers, ketones etc. decrease the surface tension of a liquid. These substances are called **surfactants**, which exhibit positive surface activity. On the other hand, substances like inorganic salts, salts of organic acids, sugars etc. increase the surface tension of a liquid and they possess negative surface activity.

3. The surface tension of protein solution is found to be altered by pH.

3.5.3 Determination of Surface Tension of Liquids by Capillary Rise Method

When a capillary tube is dipped vertically into a liquid, the liquid will rise in the tube. The process of rise or depression of a liquid in a narrow tube is called **capillarity**.

When a capillary tube with an internal radius of r is dipped vertically in a liquid having a density P, then the liquid rises in

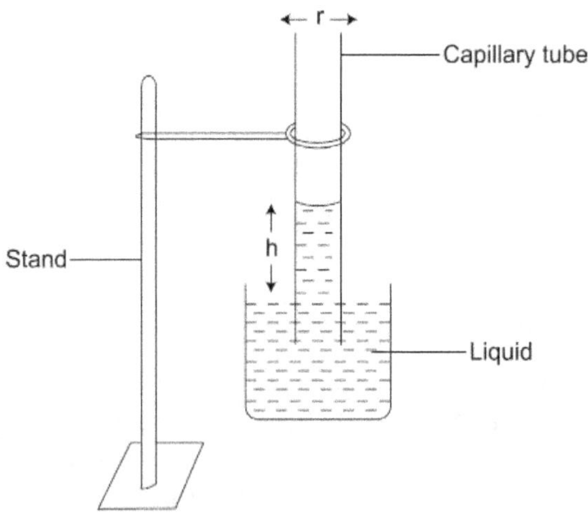

Figure 3.9. Capillary rise method to determine surface tension of a liquid.

the tube to a height as shown in Figure 3.9. The surface tension of the liquid can be calculated using the following formula:

$$\text{Surface tenstion } (T) = \frac{hPgr}{2} \text{ dynes/cm}$$

where,

  g = acceleration due to gravity.

3.5.4 Determination of Surface Tension of Liquids by Drop Weight Method

The liquid whose surface tension is to be determined is taken in the burette. With the help of pinchcock, fully formed liquid drops are made to detach themselves from the burette. The drops are collected (50 drops) in a beaker, the weight of which is already known (Figure 3.10). Then the weight of the beaker

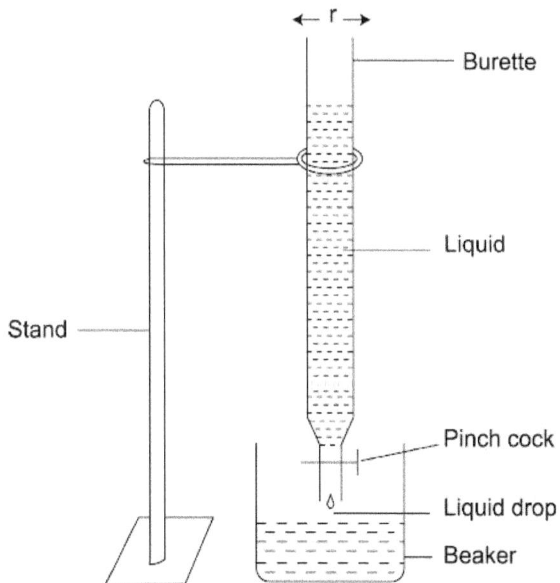

Figure 3.10. Drop weight method to determine surface tension of a liquid.

with the collected liquid is determined. From this, the mass of each drop (m) is found and the internal radius of the burette (r) is measured. The surface tension of the liquid is determined by using the formula,

$$\text{Surface tension } (T) = \frac{mg}{3.8 \times r} \text{dynes/cm}$$

where,

G = gravitational force (980) and

3.8 = A constant

3.5.5 Determination of Interfacial Surface Tension of Liquids

A similar experiment can be used to determine the interfacial surface tension of two liquids namely water and oil. Here, one end of the glass tube is connected to a bottle containing water while another end is immersed in a weighed beaker containing oil. Specific number of water drops is collected in the beaker and is weighed again. By knowing the mass of a single drop of water, the interfacial surface tension of water with the oil can be known by using the following formula:

$$\text{Interfacial surface tension } = \frac{mg \, (i - p)}{3.8 \times r} \text{dynes/cm}$$

where,

m = mass of single drop

g = gravitational force (980)

i = density of water

p = density of oil

r = radius of tube and

3.8 = a constant

3.5.6 Biological Significance of Surface Tension

- The property of molecules either to concentrate at the surface or at the interior helps to form the limiting membranes of living systems. Thus the substances which lower the surface tension concentrate at the surface layer while the substances which increase the surface tension accumulate in the interior of the liquid (**Gibbs–Thomson principle**)

- The substances which decrease the surface tension are useful in emulsification.

- Surfactants play an important role in the functioning of lungs.

- Surface tension results in efficient adsorption and thus plays a role in enzyme reactions, in formation of compounds in protoplasm and in the action of drugs and poison.

3.6 ADSORPTION

Adsorption is an important phenomenon of surfaces and causes the occurrence of higher concentration of substances at the surface of a solid or liquid than inside. Thus the concentration of molecules or ions on the surface of liquids and solids is called **adsorption**. The substance which adheres to the surface, is the **adsorbate** and the substance to which it is attached is the **adsorbent**. The forces of attraction between the adsorbate and adsorbent include dispersion forces, orientation forces and induction forces. The ionic and covalent bonds are also involved in adsorption phenomenon in which strong chemical bonds are formed between the adsorbate and the surface of the adsorbent. This involvement of bonds in adsorption is called **chemical adsorption**. Within atoms, the electrons are in continuous motion in their orbits so that the electromagnetic force outside the substances is constantly changed. As a result, the electrons of nonpolar molecules in adjacent solutions may be influenced resulting in fluctuating dipoles. When such dipoles are brought together, they orient in the same direction producing attractive forces called **dispersion forces. Orientation forces** are produced by polar molecules which have constant electrical charges on their surfaces and are stronger than the dispersion forces. **Inductive forces** are due to electrostatic forces by which the electrons are influenced from one atom to another. When the adsorbate molecules approach the adsorbent surface, repulsive forces oppose the above forces until equilibrium is attained.

An adsorbent can adsorb either the solute or the solvent in a solution depending on the nature of adsorbent and of the solution. However, adsorption of solvents occurs only rarely. The adsorption of solutes from a solution by the adsorbent is called **positive adsorption**, which decreases the concentration of the solution. On the other hand, the adsorption of solvent from the solution by the adsorbent is called **negative adsorption**, which increases the concentration of the solution.

3.6.1 Factors Affecting Adsorption

1. Adsorption is a reversible and very quick process and the extent of adsorption depends on the nature of adsorbing agents, the substances adsorbed, surface area of adsorbents and surface charges.

2. While physical adsorption occurs at low temperatures, chemical adsorption occurs only at high temperatures. Temperature alters the nature of adsorbing surfaces.

3. Adsorption is complete in more dilute solutions.

4. The energy provided by surface tension determines the adsorption.

3.6.2 Adsorption of Ions by Solids

Adsorption of ions on a solid occurs through ion exchange involving chemical and physical processes. The substances with particle diameter of less than 20 μ are mainly involved in ion exchange. The exchangeable cations include Ca^{++}, Mg^{++}, H^+, K^+, Na^+, etc. The cation exchange increases with the increase in pH and the reaction is reversible. **Cation exchange capacity (CEC)** denotes the quantity of exchangeable cations per unit weight of solid particles. The exchangeable anions include $H_2PO_4^-$, HPO_4^-, Cl^-, SO_4^-, NO_3^-, etc. Contrary to cation exchange, anion exchange increases with decrease in pH. In general, anion exchange decreases with increasing salt concentrations.

3.6.3 Adsorption of Gases by Solids

When solids adsorb gases, there is an excess concentration of gases at the surface of the solids. The best examples are charcoal, silica gel, alumina, metals like platinum, etc. The adsorption of a gas by a solid is increased with increased pressure and decreased temperature. The variation of gas adsorption under the influence of pressure and temperature can be explained by the term **adsorption isotherm.**

That is,

$$a = KP^n$$

where,

a = amount of gas adsorbed by unit mass

K = constant for the given gas

P = pressure and

n = constant for adsorbent

3.6.4 Adsorption of Ions by Liquids

The phenomenon of adsorption of ions at the surface of a solution can be explained by **Gibb's adsorption equation.** For a dilute solution of concentration 'C' and surface tension 'r', the equation is

$$S = \frac{C}{RT} \frac{dr}{dc}$$

where,

S = excess contraction of solute/square cm.

R = gas constant

T = absolute temperature

dr/dc = rate of increase of surface tension of the solution

From the above equation, it is clear that dr/dc will become negative and S, positive when a solute decreases the surface tension of a solvent. In other words, it is indicative of higher concentration of solutes at the surface than in the bulk of the solution. When any substance increases the surface tension of the solvent, the dr/dc will become positive and S, negative indicating lesser concentration of solute at the surface than in the bulk of the solution.

3.6.5 Adsorption and Colloids

The surface adsorption is higher in colloids as they provide great surface area/unit weight due to the presence of large number of particles. The particles of a colloidal system possess certain "fields of forces", which are most probably electromagnetic. These fields of force are relatively free on the surface of the particles when compared to those, which are present inside the particles. While the atoms inside the particles neutralize the forces, the forces on the surface are partially neutralized. Thus the surface of colloidal particles has free fields of force so that ions are adsorbed onto the surface. This adsorption along with the ionization of ionizable groups in the colloidal substance is responsible for the electrical properties of colloids.

On the outer surface of each colloidal particle, static charged electric system occurs in addition to adsorption of ions. In the absence of primary ionization, the electrical charge on the colloidal particle is mainly produced by the adsorption of ions. The adsorption of ions by the colloidal particles results in the formation of two layers namely an immobile layer and a mobile layer. This ionic envelop on the colloidal particle is called **Helmholtz–Gouy electrical double layer** (Figure 3.11). This double layer causes three types of potentials on the surface of the colloid particle namely **epsilon potential** or **electrochemical potential** which is the potential across all the ionic layers on the surface of colloidal particle, **zeta potential** which is the potential between the immobile layer and mobile layer and **stern potential** which is the potential between the immobile layer and the surface of the particle (Figure 3.12). Among them, electrochemical potential determines the electrical properties of colloidal particles. In many colloidal substances, this potential measures about 30 millivolt, when water is the dispersion medium. In this way, on passing electrical current through a colloidal solution (Electrophoresis) all colloidal particles possess like electrical charges so that they move either towards the anode or cathode. If this potential is lowered (less than 15 mV, then the colloid particle attains an 'isoelectric point' where there is no electrophoresis.

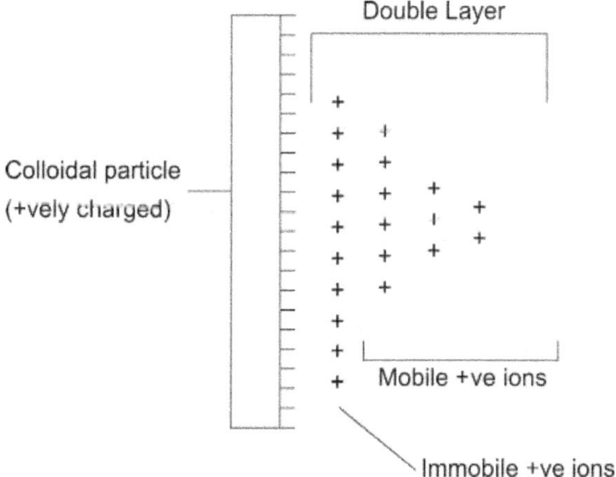

Figure 3.11. Helmholtz–Gouy double layer.

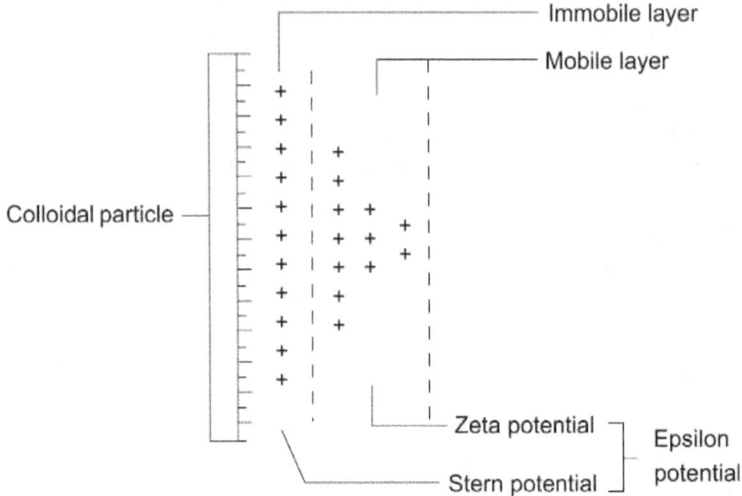

Figure 3.12. Potentials on double layer.

3.6.6 Biological Significance of Adsorption

- Adsorption is responsible for the orientation of macromolecules in cytoplasmic membranes. The structural integration and normal functioning of the cell is mainly due to the molecular association between proteins, carbohydrates and salts, which in turn depends on adsorption.

- The biological membrane with colloidal molecules provides large surface area required by enzyme reactions so that adsorption increases the chemical reactions in biological systems.

- Adsorption helps the enzymes to combine with substances and to purify enzymes.

- It is useful in the emulsification of liquids through the adsorption of emulsifiers.

3.7 HYDROTROPY

Water is the universal intracellular and extracellular medium and has two important features namely possession of hydrogen bonds and dipole nature. The substances, which take advantage of these properties, are readily soluble in water and are called **hydrophilic substances.** The molecules, which have hydrogen-bonding groups, are capable of combining with water by hydrogen bonds. Thus hydrogen bond acceptors or donors dissolve readily in water. Therefore, water readily dissolves hydroxyl compounds, amines, sulphydryl compounds, esters, ketones and a number of other organic compounds. Water is also an excellent solvent for ionic compounds like sodium chloride due to its dipole nature. In an aqueous solution, the water dipoles interact with cations and anions so that the ions become hydrated. In other words, shells of water molecules called **hydration shells** surround the ions. The dipole nature of the water molecule also makes its ability to dissolve non-ionic but polar organic molecules like phenols, esters and amides. As these molecules have large dipole moments, the interaction of these molecules with water dipole increases their solubility in water. In polar substances, there is an uneven distribution of charges within the molecules. This makes the polar compounds to dissolve in water.

Substances like hydrocarbons, which are nonpolar, non-ionic and incapable of hydrogen bonding exhibit no water solubility or only limited solubility in water. These substances are called **hydrophobic substances**. When hydrophobic molecules dissolve in water, hydration shells are not formed but the water lattice forms 'cages' called **clathrate structures**. The nonpolar molecules like lipids are repelled by water and they usually group together in the presence of water. Many proteins containing high salt concentrations or having strong ionic strength become insoluble in water. Thus solubility in concentrated solutions varies with ionic strength. Triglycerides are nonpolar and do not form hydrogen bonds with water molecules and so they are not soluble in water. Phospholipid molecules consist of phosphate head groups and two hydrocarbon tails from fatty acids. The phosphate head of a phospholipid molecule possesses an electrical charge which makes it soluble in water while the tail is insoluble. As the fatty acids have hydrocarbon tails, they are insoluble in water.

On the contrary, **amphipathic substances** possess both hydrophilic and hydrophobic properties as they have hydrophilic head groups and hydrophobic tails (hydrocarbons). When these molecules are dissolved in water, they may form a monolayer on the water surface. If the mixture is vigorously stirred, micelles or bilayer vesicles are formed. Thus amphipathic molecules are responsible for the formation of membrane bilayers in biological systems.

The process by which the water-insoluble substances are made into water-soluble substances is called **hydrotropy.** The substances which have the ability to convert water-insoluble substances into water soluble ones, are called hydrotropic substances. The examples for hydrotropic substances include cholic acid, benzoic acid, phenylacetic acid, hippuric acid, and soaps of higher fatty acids and so on. Cholic acid, deoxycholic acid, lithocholic acid and chenodeoxycholic acid are different bile acids. Substances like fatty acids, phenol, etc. combine with deoxycholic acid to form choleic acids, which are water-soluble. As a result, insoluble fatty acids, cholesterol and fat-soluble vitamins are made soluble in water. A bile salt molecule, which is formed by bile acid, is amphipathic in nature and forms micelles through emulsification of lipids. These micelles are acted on by water-soluble enzymes, thus facilitating the absorption of lipids in intestinal cells. In this way, the bile salt is an emulsifying agent, which makes fats miscible with water.

3.7.1 Biological Importance of Hydrotropy

- Hydrotropy is useful to dissolve substances which can easily diffuse through the membranes.

- Hydrotropic substances are found in bile, intestinal juice, blood plasma, etc. where they are useful in absorption of insoluble substances such as cholesterol, fatty acids and so on.

- Hydrotropic substances reduce the surface tension of liquids, thus helping in the emulsification of fat.

- When hydrotrophic substances come in contact with water, water molecules surround ions and polar groups and ions or molecules are dissociated from each other. In other words, water dissociates ions or molecules, which move freely in solution so that the solution becomes more reactive chemically.

3.8 PRECIPITATION

In an atom, the nucleus is positively charged and is surrounded by negatively charged electrons. An atom in solution is surrounded by the particles of the solution and so a double layer is formed around the nucleus namely an inner immobile layer and an outer mobile layer. The net potential between these two layers is called as zeta potential or electrokinetic potential which is due to electrostatic and adsorptive interactions. The stability of a solution depends on zeta potential of its particles. When electrolytes are added to a solution, zeta potential of the particles are reduced with the resultant neutralization of charges and breakage of double layer leading to the precipitation of the solution.

The colloidal particles have a diameter ranging from 1 mµ to 200 mµ so that they are intermediates between molecules and substances. These are electrically charged. In order to bring precipitation in colloidal systems, their electrical charges must be neutralized by the ions of opposite charge, thus bringing the particles to isoelectric state. Using colloidal system of opposite electrical charge can precipitate both lyophilic and lyophobic colloids. Thus the stability of a solution depends on the charges of particles. This explains various precipitation reactions in living systems. For example, calcium ions reduce the zeta potential of dispersed particles during blood coagulation so that they become aggregated to form a clot due to the breakage of the double layer in the particles.

Two major factors determine the stability of both colloidal systems namely the charge of colloidal particles and the degree of solvation power. The lyophilic colloids have high degree of solvation of particles so that the neutralization of charges alone cannot bring precipitation. Therefore, they require addition of alcohol or strong solutions to bring precipitation. However, the lyophilic colloids can be precipitated by the addition of a large amount of soluble solutes (**salting out processes**). For example, addition of alcohol causes precipitation of serum proteins. On the other hand, the hydrophobic colloidal particles have like electrical charges, so they repel each other. Some of the colloidal particles are adsorbed by the electrolyte ions, which neutralize the charges. Therefore, hydrophobic colloids can easily be precipitated by addition of a small quantity of electrolytes.

In addition to the electrolytes, H^+ and OH^- ions, heat, ultraviolet rays, ultrasonic sound, acids, alkalies, etc. also alter the zeta potential of the colloidal particles resulting in precipitation. Moreover, precipitation of solutions can also be brought by mutual cancellation of charges by the addition of a negative substance to a positive one or vice versa (**mutual precipitation**).

3.8.1 Biological Significance of Precipitation

In living organisms, physiological colloidal systems are mostly lyophobic in nature. Therefore, they can tolerate alterations in concentration of inorganic ions in the body.

3.9 VISCOSITY

Viscosity is one of the transport properties characteristic of all liquids and gases. It can be defined as the resistance of a liquid to the flow due to the frictional effect of one layer of the liquid over another. In other words, the difference in the rate of flow of liquids or gases is called viscosity which is due to the resistance formed when a layer of liquid flows over another layer. The term fluidity refers to the tendency of a liquid to flow whereas viscosity is the measure of resistance offered to the flow of a liquid. The liquids with low viscosity are called **mobile liquids** while those which have high viscosity are called **viscous fluids**. In general, the addition of macromolecules to a solvent increases its viscosity. As the body fluids contain a number of macromolecules, they are viscous in nature. The ratio of solution to solvent viscosity is called relative viscosity and the ratio of the change in viscosity in a solvent is called specific viscosity.

When a fluid flows in one direction with a specific velocity, it is broken up into thin layers each of which moves with different velocity. If we consider only two layers, the upper layer moves faster than the lower layer, which slows down the movement of upper layer. At the same time, the fast-moving upper layer speeds up the movement of the lower layer. To keep the velocities of the two layers unchanged, a force is necessary. This force is called **shear stress** (S), which must be proportional to the velocity gradient.

In the flow of a fluid through a glass tube with a non-uniform linear velocity (v), one layer of the liquid experiences resistance when it is flowing over another layer. Suppose two layers are separated by a distance x and a shear stress is applied, then each layer is moved with different velocity as shown in Figure 3.13. The velocity gradient of the fluid layers can be related with the shear stress by the equation,

$$\frac{F}{A} = \eta \, \frac{dv}{dx}$$

where,

F/A = shear stress

η = viscosity coefficient of the liquid and

dv/dx = difference in viscosity between two layers separated by a distance *dx*

The viscosity coefficient is expressed in dynes/cm/sec., which is usually called as poise. The velocity of the flow of a liquid depends on its viscosity coefficient. If the velocity of the flow is lesser than the critical velocity in a fluid then the fluid flows in an orderly manner and is called **steady flow** or **stream-line flow**. On the other hand, if the velocity of the flow is more than the critical velocity, the liquid flow stops and becomes unsteady.

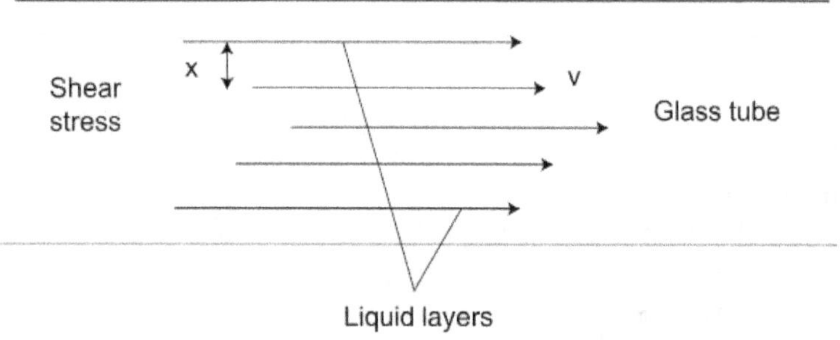

Figure 3.13. Velocity gradient in fluid layers.

3.9.1 Factors Affecting Viscosity

1. Viscosity of a liquid decreases as the temperature increases.

2. At an isoelectric pH, the viscosity of the solution decreases and when the pH is raised or lowered, the viscosity increases.

3. Viscosity is directly proportional to the pressure in liquids.

4. Viscosity of a liquid depends on size, shape and fluctuation of its particles.

3.9.2 Determination of Viscosity of Liquids

Capillary flow method   This method is useful to determine the viscosity of low viscous fluids. An aspiratory chamber is filled with water, the mouth of which is tightly closed with a single-holed rubber stopper. One end of a rubber tube with a pinch cock is fitted to the hole of the chamber and another end to a capillary tube of known length. The capillary tube is made horizontal and the flow of water in the tube is controlled by the pinch cock. An empty beaker of known weight is placed under the capillary tube (Figure 3.4). The pinch cock is slowly released, starting a stopwatch simultaneously. The water from the capillary tube is allowed to get collected in the beaker for a specific time period (about 2 minutes). The beaker with water

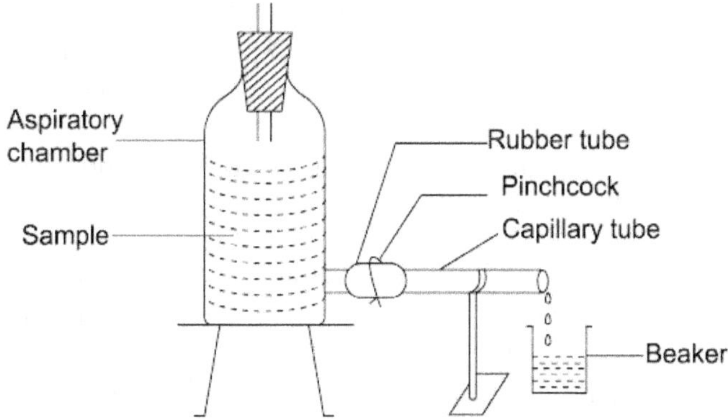

Figure 3.14 Capillary flow method to determine viscosity of a liquid.

is weighed. The initial and final water levels in the chamber are recorded and subtracting the final water level from the initial water level arrives at the value h. The radius of the capillary tube is known and the viscosity of the liquid is calculated using the formula.

$$\eta = \frac{\pi\, pgr^4}{8l} \times \frac{ht}{m}$$

where,

$\pi = 3.14$

p = density of water

g = gravity of water

r = radius of the capillary tube

l = length of the capillary tube

8 = constant

h = height of water in the chamber

t = time taken and

m = mass of water

Stokes method   This method is useful to determine the viscosity of highly viscous fluids. One end of a wide glass pipette is closed with a rubber stopper and is filled with the given liquid. Two threads are tied in the pipette at a finite distance (Figure 3.15). A small steel ball is taken and its radius is measured with the help of a screw gauge. The steel ball is now dropped into the liquid contained in the pipette, simultaneously switching on a stop watch. The time taken for the ball to travel a length in the liquid is noted. The experiment is repeated by using three or four balls of different diameters. The viscosity of the fluid is determined by using the formula,

$$\eta = \frac{2}{9}g(\sigma - p)\frac{a^2}{v}$$

where,

g = gravity of the liquid

σ = density of the liquid

p = density of the steel ball

a = radius of the steel ball

v = time taken to travel and

9 = a constant.

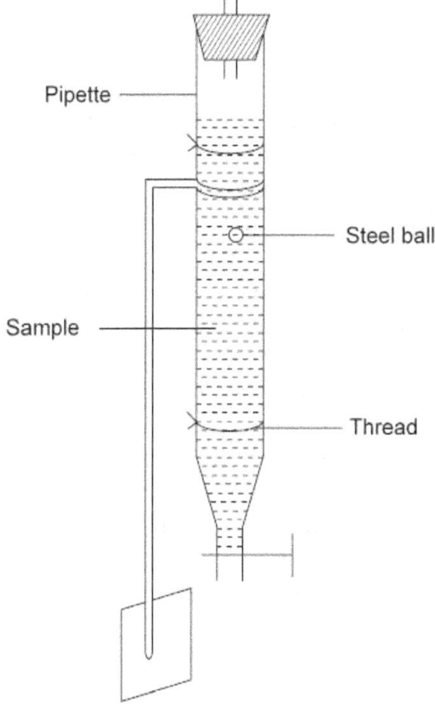

Figure 3.15. Stokes method to determine viscosity of a liquid.

3.9.3 Biological Significance of Viscosity

- The blood of man is highly viscous and so that even small alterations in viscosity will affect the heart beat.

- The amoeboid movement in living cells is mainly due to changes in the viscosity of cytoplasm.

- Viscometry is useful in the study of properties of macromolecules such as asymmetry, size and shape of macromolecules, protein denaturation, and chemical modification in protein molecules, aggregation and dissociation of proteins, interactions between actin and myosin, intrastrand disulphide bonds in proteins, polymerization of DNA and so on.

3.10 COLLOIDS

A colloid is a substance in which the particles are dispersed in a dispersion medium and is an intermediate state between a suspension and a solution. A colloidal system is thus a heterogeneous diphase system consisting of a dispersion phase constituting the bulk and the dispersed phase. The dispersed phase includes macromolecules such as proteins, nucleic acids, etc. The dispersed particles are otherwise called as micelles. The dispersion phase includes the medium in which the insoluble colloidal particles are dispersed. The dispersed particles are otherwise called as inter-micellar particles. The combination of any two phases can form a colloid of different types namely gas/liquid, gas/solid, liquid/gas, liquid/liquid, liquid/solid, solid/gas, solid/liquid and solid/solid. Solid/liquid colloid is called as a **sol** whereas; liquid/liquid colloid is called as an **emulsion** which contains two immiscible liquids. Sols in which water is the dispersion medium are called **hydrosols** whereas sols in which alcohol is the dispersion medium are called **alcohols**.

3.10.1 Types of Colloids

As a colloidal system consist of two phases namely dispersed phase made up of colloidal particles and dispersion medium made up of solvent, many types of combinations of these two phases are possible and are shown in Table 3.3.

Table 3.3. Combinations of two phases in colloidal systems.

Dispersed phase	Dispersion medium	Example
Solid	Solid	Alloys
Solid	Liquid	Starch in water
Solid	Gas	Dust in air
Liquid	Solid	Butter and cheese
Liquid	Liquid	Oil in water
Liquid	Gas	Clouds
Gas	Solid	Pumice stone
Gas	Liquid	Froth on beer

In general, the colloids are classified into two major groups namely **lyophilic colloids** and **lyophobic colloids** based on the interactions between different phases. The substances like proteins nucleic acids, starches, etc. are the lyophilic colloids in which the solvent and the particles are highly attracted. The lyophilic colloid includes hydrosols and alcohols or emulsoids. The emulsoids have a high degree of solvation of the particles that they cannot be precipitated easily but can be precipitated only by highly concentrated electrolytes. The lyophobic colloids are unstable and 'solvent hating substances' and are also called as **suspensoids** which can be easily precipitated. In addition to the above colloids, spontaneously aggregated macromolecules formed by the micelles also occur and are called **association colloids**, which are highly stable.

3.10.2 Characteristics of Colloids

Kinetic properties

1. In a colloidal system, the size of the particles range from $1\,\mu$ to $0.1\,\mu$ in diameter and the particles are formed either by aggregation of small molecules or by disintegration of large molecules. The colloidal state of matter is intermediate between molecules and suspensions. The colloidal particles can be seen only under the ultra microscope. As the colloidal particles are very small in size, they can pass through ordinary filter paper but cannot pass through animal membranes.

2. A parchment membrane can retain the colloidal particles and this property is employed in the process of dialysis.

3. The colloidal systems exhibit lesser osmotic pressure when compared to true solutions.

Optical properties

1. The colloidal particles are capable of scattering light and this phenomenon is called **Tyndal effect,** which is applied in the construction of ultramicroscope.

2. The colloidal particles exhibit a random zigzag motion called **Brownian movement**, which is due to the bombardment of the colloidal particles, by the molecules of the dispersion medium.

Electrical properties   Most of the biopolymers in solution are colloidal in nature and are electrically charged. Their charges may be strong or weak; their distribution may be symmetric or asymmetric and their origin may be external or internal. For example, a spherical colloidal particle with a negatively charged surface will attract the positive ions from the dispersion medium. Therefore, the particle will be surrounded by a region of ions having opposite charge. This condition is called Helmholtz's double layer. According to electrical double layer theory, the ions of opposite charge would concentrate and orient as an immobile layer of a colloidal particle. Here the potential depends on the charge of the particle and its radius as well as the dielectric constant of the medium. However, this potential is reduced by the outer layer of the oppositely charged ions (mobile layer or diffuse layer) (Figure 3.16). Therefore, the net potential between the immobile layer and the mobile layer is called zeta potential or electrokinetic potential. In addition to zeta potential, the charged layer of colloidal particles provides other potentials also. The immobile layer, which is in contact with colloidal particles with oppositely charged ions, is otherwise called as stern layer. Thus the potential between stern layer and surface of the particle is called stern potential. Moreover, the potential difference between the surface of the particle and all the ionic layers in solution around the particles is called electrochemical potential (E).

$$E = \text{Zeta potential} + \text{stern potential.}$$

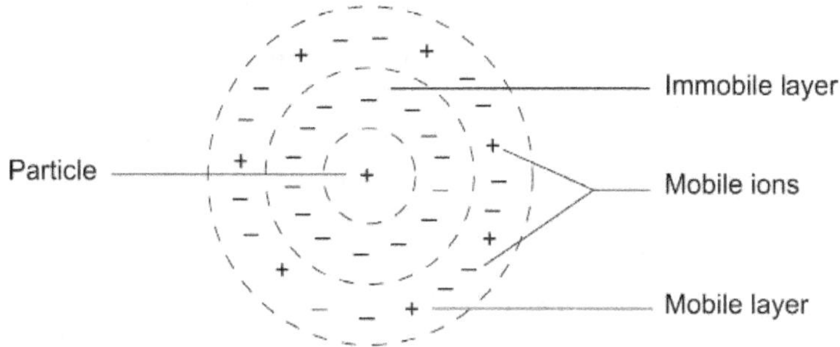

Figure 3.16. Electrical double layer around a colloidal particle.

The addition of electrolytes neutralizes the charges in the colloidal system thereby collapsing the double layer.

3.10.3 Stability of Colloids

The stability of both lyophilic and lyophobic colloids is due to like charges and force of repulsion between individual particles. Moreover, solvation or hydration of the particles in lyophilic colloids is associated with the solvent molecule.

3.10.4 Gel

Coagulation of a solution results in the formation of a gel, which is a semi-rigid and jelly-like mass and can be considered as a colloidal system consisting of a liquid dispersed in solid. The elastic gels when placed in water imbibe large quantity of water and swell in size (agar and gelatin). The non-elastic gels when dried, become powdery and lose their elasticity.

3.10.5 Emulsions

These are liquid/liquid colloids in which two liquids are immiscible but either liquids can be dispersed in the other. Emulsions are in general unstable but the emulsifying agents reduce the interfacial tension between the liquids so that stability is maintained in the liquid/liquid colloid.

3.10.6 Techniques for the Separation of Colloids

Electrophoresis   When an electric current is passed through a solution, all the colloidal particles move either towards the anode or the cathode, as all the colloidal particles possess like electrical charges. For example, some like proteins in acid solution, haemoglobin, etc. have positively charged particles whereas, proteins in alkaline solution and metal solutions possess negatively charged particles. Therefore, they can be easily separated by electrophoresis. By the addition of suitable reagents, the same colloids may be charged either positively or negatively.

Electroosmosis   Here an electric current is passed through a solution in which the colloidal particles are kept stationary. As a result, the dispersion medium moves towards the electrode.

Precipitation   When the electrochemical potential of a colloid is lowered (less than 15 mV), then the charges of colloidal particles are neutralized and are brought to their iso-electric points. As a result, the colloids are precipitated. Salt solutions of sodium chloride and ammonium sulphate and acids such as trichloroacetic acid, etc. bring the charges of colloidal particles to the isoelectric point so that they are used to precipitate colloids.

3.10.7 Biological Importance of Colloids

- The colloids provide large surface area for enzyme action and adsorption.

- Proteins, phospholipids, etc. act as amphoteric colloids and react with acids and alkalies, thus helping in maintaining the blood pH.

- Colloidal proteins and phospholipids help in fat absorption and protect the bile salts from precipitation by electrolytes.

- The osmotic pressure exerted by the colloidal system plays an important role in water transport in the body and urine formation.

- The colloids imbibe the tissue fluids in glandular cells, thus producing glandular secretions.

- The colloids are responsible for selective membrane permeability as explained by Donnan membrane equilibrium and membrane hydrolysis.

3.11 GIBB'S DONNAN EQUILIBRIUM (DONNAN MEMBRANE EQUILIBRIUM)

This phenomenon explains the volume distribution of ions on either side of two compartments separated by a semipermeable membrane. If electrolytes are present in equal concentrations on either side of a freely permeable membrane, all the ions of the electrolytes will be equally distributed on either side of the membrane. If the membrane is impermeable to all the ions, then the ions will be unequally distributed on either side of the membrane. On the other hand, selective membrane permeability of certain ions is mainly due to their colloidal state. Donnan equilibrium explains this role of colloidal particles in membrane permeability.

According to this principle, the quality and quantity of non-diffusible ion or ions on one side of the membrane would influence the diffusion of the diffusible ion or ions. When a porous membrane separates two solutions with different concentrations in two ions, these two ions will diffuse across the membrane until their concentrations become equal on both sides. This ionic distribution could not occur when a larger anion, which is not able to cross through the pores of the membrane, is present on one side. For example, when potassium proteinate (potassium salt of protein) is separated from potassium chloride solution in compartments A and B by a membrane, the K^+ ions from both sides and Cl^- ions from KCl compartment will freely diffuse across the membrane until K⁺ and Cl⁻ ions are equally distributed in two compartments. On the other hand, proteinate ions will not cross the membrane resulting in an excess of negative charge on their side.

In the above systems, K⁺, Pr⁻ and Cl⁻ ions will be distributed in compartment A whereas the compartment B will contain only K⁺ and Cl⁻ ions at an equilibrium level. Therefore, Pr⁻ ions would produce an excess of negative charge in compartment A. K⁺ ions have to balance two negative ions (Pr⁻ and Cl⁻) and it has to balance only Cl⁻ ion in compartment B. Therefore the concentration of K⁺ ions in compartment A will be greater than in compartment B. So the total ionic concentration will be greater in compartment A than in compartment B. Thus the system experiences two opposing forces namely an electrical force exerted by Pr⁻ ions causing an excess quantity of K⁺ ions in compartment A and a chemical force caused by the movable ions (K⁺ and Cl⁻ ions) to become equal in concentration in both compartments. As a result, an equilibrium is attained between these two opposing forces with

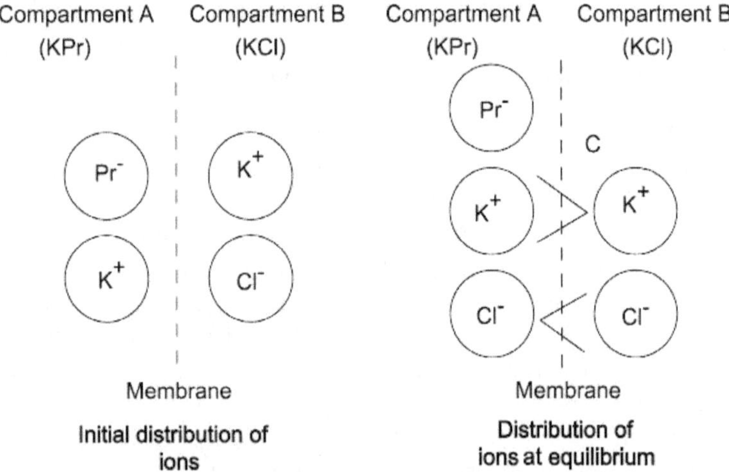

Figure 3.17. Initial and final distribution of ions in two compartments.

the resultant imbalance of ions on either side of the membrane (Figure 3.17). Thermodynamically, the concentrations of K^+ and Cl^- ions in both compartments should be similar at an equilibrium level. But the concentration of K^+ ions is greater in compartment A and the concentration of Cl^- ions in compartment B. This is due to **Donnan membrane equilibrium.**

Thus the Donnan equilibrium causes an accumulation of oppositely charged diffusible ions on a side in which non-diffusible ions are present and also accumulation of diffusible ions of the same charge on the other side of the membrane. The net result is that the total concentration of ions will be greater on the side of non-diffusible ions leading to osmotic imbalance between the two sides. However, the Donnan equilibrium is displaced when a solution containing monovalent and divalent cations is diluted. Here, the movement of divalent ions increases whereas the movement of monovalent ions decreases. That is, the ratio of divalent to monovalent ions increases. On the other hand, the ratio of monovalent to divalent ions increases when the solution concentration is increased. If the particles of a solution have higher exchange capacity, the movement of divalent ions will be higher than the monovalent ions or vice versa.

3.11.1 Donnan Equilibrium in Living Systems

1. Donnan effect causes an unequal distribution of ions on either side of the cell membrane thus producing differential osmotic pressure. A semipermeable membrane envelops the living cells. The protein molecules found in the cytoplasm of cells with negative charge cannot cross the membranes. To compensate this, smaller diffusible ions are distributed predominantly inside the cells. In other words, the transmembrane differences of the non-diffusible electrolytes bring out changes in electrochemical gradients between the inside and outside cells thus altering the distribution of diffusible ions.

2. Lesser pH of red blood cells than the plasma is mainly due to the presence of negatively charged haemoglobin in the red cells, which influence the oppositely charged H^+ ions resulting in decreased pH.

3. The osmotic imbalance due to Donnan effect is the main cause for the Donnan effect cause for inhibition of gels.

4. Donnan effect causes membrane hydrolysis in cells. One side of the cell may either be acidic or alkaline due to the

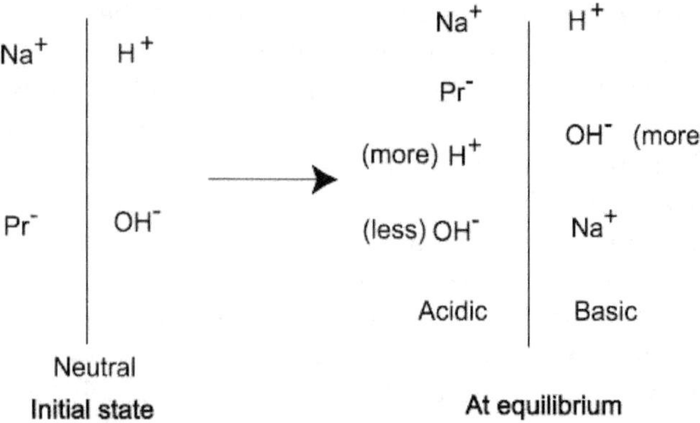

Figure 3.18. Interaction between sodium proteinate and water.

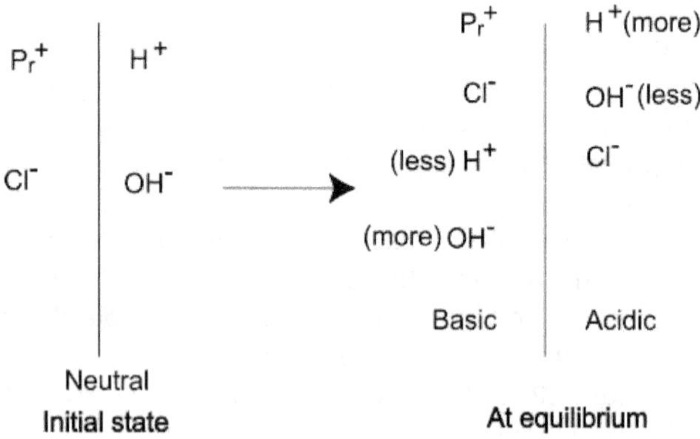

Figure 3.19. Interaction between Pr^+, Cl^- and water.

charges of non-diffusible ions so that this side of the system is concerned with H^+ and OH^- ions of water. This shifting of ions is called membrane hydrolysis. For example, in the interaction between sodium proteinate and water through a membrane, there is accumulation of H^+ ions on the side containing sodium proteinate so that this side becomes acidic (Figure 3.18).

5. In the interaction between Pr^+Cl^- and water through a membrane, there is an accumulation of OH^- ions on the side containing Pr^+Cl^- so that this side becomes basic (Figure 3.19).

REVIEW QUESTIONS

1. Explain different types of diffusion and add a note on various factors which influence diffusion.

2. Write a note on Fick's law of diffusion.

3. Define osmosis and osmotic pressure.

4. How can osmotic pressure of samples be determined?

5. Describe mercury osmometer and cryoscopic osmometer.

6. How can osmotic pressure be calculated from Vant Hoff's equation?

7. Write an essay on role of filtration in biological systems.

8. Explain the mechanism of dialysis and add a note on its role in artificial kidney.

9. What is surface tension? Explain kinetic theory of surface tension.

10. Describe a method to determine surface tension and interfacial surface tension of liquids.

11. Write an essay on adsorption.

12. Explain the mechanism of adsorption of ions by solids.

13. Explain the mechanism of adsorption of gases by solids.

14. Explain the mechanism of adsorption of colloids.

15. What is Helmholtz–Gouy double layer? Enumerate its biological significance.

16. Write an essay on hydrotropy and its role in living organisms.

17. Explain the mechanism of precipitation and its biological significance.

18. What is viscosity? Explain the various factors which affect viscosity of liquids.

19. Describe velocity gradient in fluid layers.

20. Explain the methods used to determine viscosity of liquids.

21. Classify colloids.

22. Give an account of the characteristics of colloids.

23. Explain the electrical double layer around a colloidal particle.

24. What are the techniques used to separate colloidal particles?

25. What is Donnan membrane equilibrium? Explain it with an example.

26. What is the significance of Donnan membrane equilibrium in living systems?

4

PRINCIPLES OF OPTICS IN BIOLOGICAL STUDIES

4.1 CHARACTERISTICS OF LIGHT

Light is one of the most vital and essential factors for the existence of life. Light can be defined as the visible part of the spectrum of solar radiant energy. The major natural sources of light includes sunlight, moonlight, starlight and light from luminescent organisms. Light spreads out in the form of divergent rays from a point of source. The light rays occur as narrow bundles and so light can be considered to consist of waves. The height of a light wave is called **amplitude** and the length of one full wave from one point to another is called **wave length** (λ). The number of oscillations of light waves per second is called **frequency** (b). When light passes through a medium, the light rays are separated according to their wavelengths and this phenomenon is called **chromatic dispersion**. The refractive index of light depends on its wavelength. In free space, the velocity of light is the highest as it encounters no atoms. When light passes through matter, it interacts with the electrons of atoms so that its velocity is lesser. Thus the velocity of light is inversely proportional to the density of atoms.

Reflection and refraction of light are the two important phenomena involved in geometrical optics. When light passes from one medium to another, a part of the incident light is reflected and a part is refracted or deviated from its initial direction. When the light rays are passed through a prism, light becomes dispersed and the dispersed rays come out and are visible to the naked eye (**Visible spectrum**) as shown in Figure 4.1.

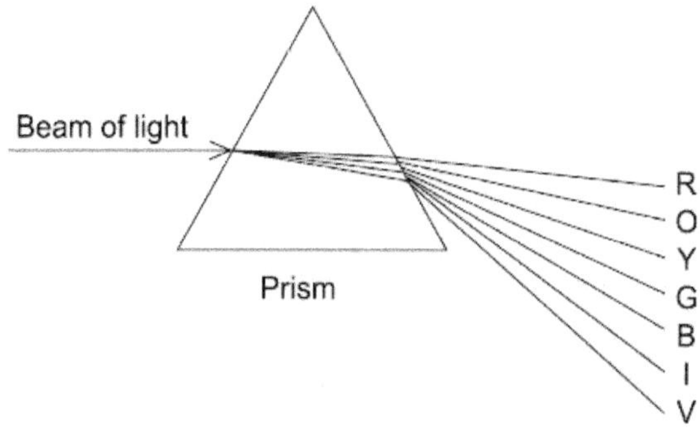

Figure 4.1. Dispersion of light.

When the light rays are passed through a slit, some rays emerge out in straight lines through the slit and are called **direct** or **non-deviated rays.** Some rays bend sideways undergoing deviations from the original path and are called **deviated** or **diffracted rays** as in Figure 4.2.

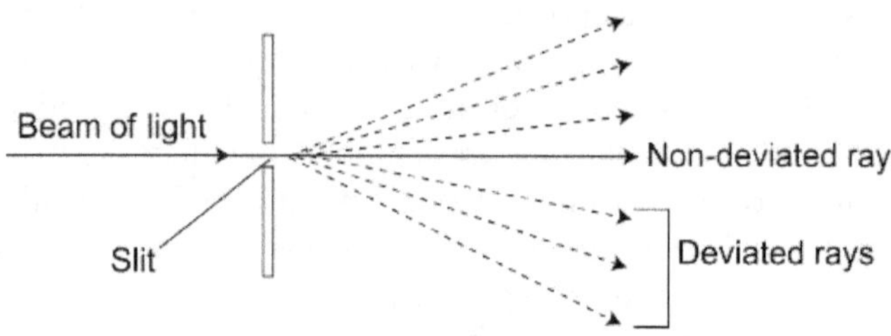

Figure 4.2. Diffraction of light.

When a light ray travels from one medium to another, the speed of light undergoes a change and is called **refraction**. As the speed of light is different in different media, refraction occurs. In other words, the magnitude of refraction depends on the speed of light in each medium. It means that the velocity of light changes as it travels from one medium to another. But the velocity of light in a given medium depends on the wavelength or frequency. Light with a single wavelength is called **monochromatic light**. The ratio of velocities of light in free space and in a medium forms the **refractive index** of the medium. Suppose, c and v are the velocities of light in free space and in a medium respectively, then the ratio c/v is the refractive index of the medium (μ).

$$\mu = \frac{c}{v}$$

But, $$c = f\lambda$$

where,

f = frequency of vibration and

λ = wavelength

When light passes from one medium to another, the frequency remains unaltered but the wavelength is different in different media. For example, if λ and λ_m are wavelengths in free space and in the medium, then

$$\mu = \frac{\lambda}{\lambda_m}$$

The principle of dispersion, diffraction, reflection and refraction of light are used in designing microscopes.

MICROSCOPY

Microscopy involves the use of optical principles in instruments called microscopes, which are of different types. All microscopes are magnifying devices and are used to have an enlarged image of the object and a contrast between the object and the background. They are characterized by **magnification** and **resolving power.** Magnification power of a microscope can be calculated by the formula,

$$\text{Magnification } (M) = \frac{10}{f} + 1$$

where,

10 = Distance of vision and

f = Focal length of the lens

The resolving power of a microscope denotes its ability to differentiate two objects and it increases with the decrease in wavelength of illumination. The objective lens determines the resolving power of a microscope and it can be calculated by using the following formula:

$$\text{Resolving power } (R) = \frac{0.61\lambda}{NA}$$

where,

λ = Wavelength of illumination

NA = Numerical aperture of the objective lens and

0.61 = A constant

TYPES OF MICROSCOPES

4.2 COMPOUND MICROSCOPE

The optical principle of a compound microscope is shown in Figure 4.3. It has the following components:

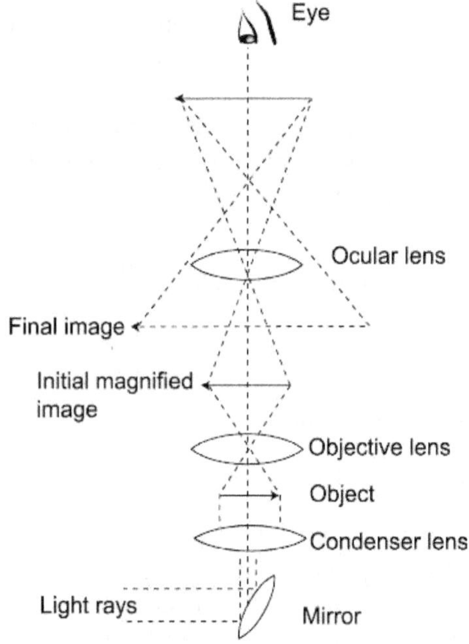

Figure 4.3. Optical principle of a compound microscope.

Light source   Ordinary sunlight or electrical light is the source of illumination and is reflected to the object through a mirror (Plano-concave mirror)

Condenser lens system   The condenser collects and focuses the light on the specimen.

Objective lens   It brings initial magnification of the specimen and produces an inverted and magnified image. Three objective lenses with different magnifying powers are provided to a revolving nose piece for desired magnification.

Oil immersion objective   When a beam of light is passed from air into glass, it becomes deviated but retains its original position passing back from glass into air. This property of light has only little effect on low power objectives. When high power objectives (50 X and 100 X) are used, light rays become bent such that the amount of light and the resolving power of the objective lens are affected. This problem is solved by placing oil which has the same optical properties as glass (cedar wood oil) between the specimen and lens. As a result, the light passes in a straight line from glass through the oil and back to the glass as shown in Figure 4.4.

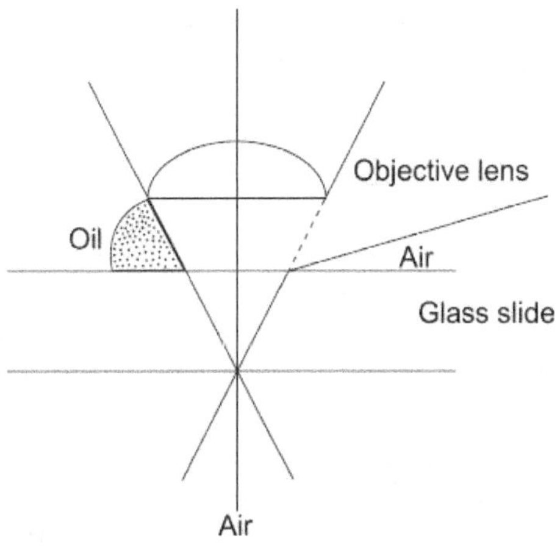

Figure 4.4. Passage of light through air and oil.

Eyepiece or ocular lens   It brings further magnification of the image formed by the objective lens and focuses this magnified image to the eye or photographic plate.

Focusing knobs   A pair of focusing knobs for coarse and fine adjustments is provided in order to adjust the body tube for correct focusing.

Mechanical stage   A mechanical stage with a pair of clips is present between the objective lens and condenser. It helps to hold the glass slide on which the specimen is placed and to bring the specimen to the field of vision.

Iris diaphragm   It is present in between the stage and mirror to control the intensity of light.

Base   The entire set-up is mounted on a solid, strong and heavy base with an arm which can be inclined to any convenient angle.

All compound microscopes are designed in a standard pattern. The maximum limit of the resolution power of a compound microscope is about 1500 times.

4.3 PHASE CONTRAST MICROSCOPE

The light rays travel in a straight line as waves. If two such rays having similar waves travel together, they are said to be in phase in which the two rays help each other producing bright illumination. If two such waves are out of step with each other (out of phase), they interfere and hinder with each other producing retarded illumination. When a light ray is passed through an object, which has a high refractive index than the medium, then there is a delay or retardation of light waves. This is called **phase change.** The biological structures exhibit phase changes due to differences in refractive index and thickness of different parts. But these phase changes are smaller and cannot be detected by a compound microscope.

Phase contrast microscope amplifies these small phase changes, which can be detected by the eye or by a photoelectric plate.

4.3.1 Optical Principle

The optical principle of phase contrast microscope is the same as that of compound microscope but it differs in the provision

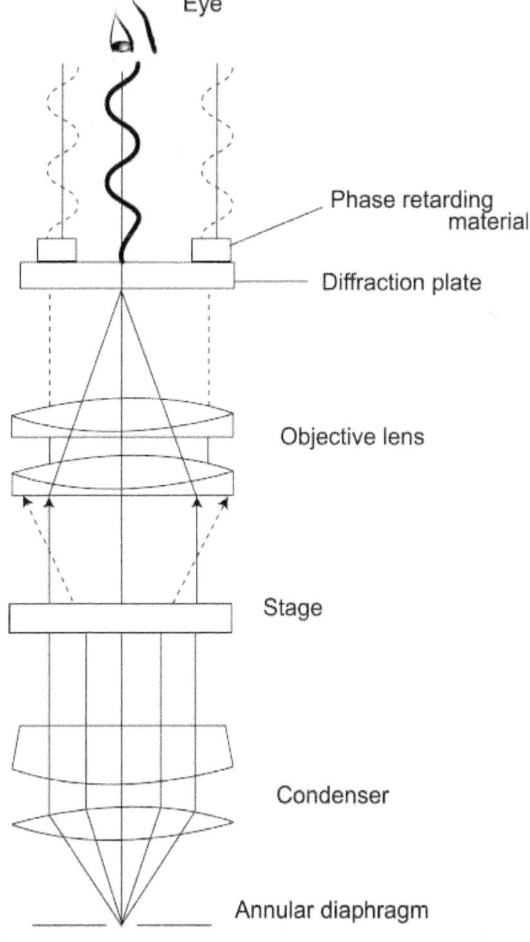

Figure 4.5. Optical principle of phase contrast microscope.

of a substage annular diaphragm and a diffraction plate in the objective lens as shown in the figure 4.5. The substage annular diaphragm provides a narrow illumination which is focused on the object.

When the light is passed through the lens, some rays travel in straight lines and others deviate laterally. The diffraction plates in the objective lens separate the deviated and non-deviated light rays coming from the object. The phase retarding material placed on the diffraction plate is used to change the phase of deviated and non-deviated rays. In a bright medium, the deviated and non-deviated rays are added and the object appears brighter than the surrounding. In a dark medium, the two types of rays hinder each other so that the object appears darker than the surrounding.

Uses   Phase contrast microscope is very useful in *in vivo* studies. i.e., for the observation of living tissues (cell division, phagocytosis, pinocytosis, etc.), bacteria and flagellates. Transparent microorganisms suspended in a fluid medium can also be examined under the phase contrast microscope.

4.4 INTERFERENCE MICROSCOPE

The principle of interference microscope is similar to that of phase contrast microscope and its optical principle is shown in Figure 4.6.

Here, the light from the source is split up into two beams of which one is passed through the object and another one is made to bypass the object. When the two beams are allowed to recombine, they interfere with each other. By using a filter, the amplitude and phase of the beams can be altered. Here, the beam, which has crossed the object, exhibits retardation when compared to the direct beam. In other words, it has undergone a phase change which is due to differential thickness of the object (t) and the refractive indices of the object (n_o) and of its

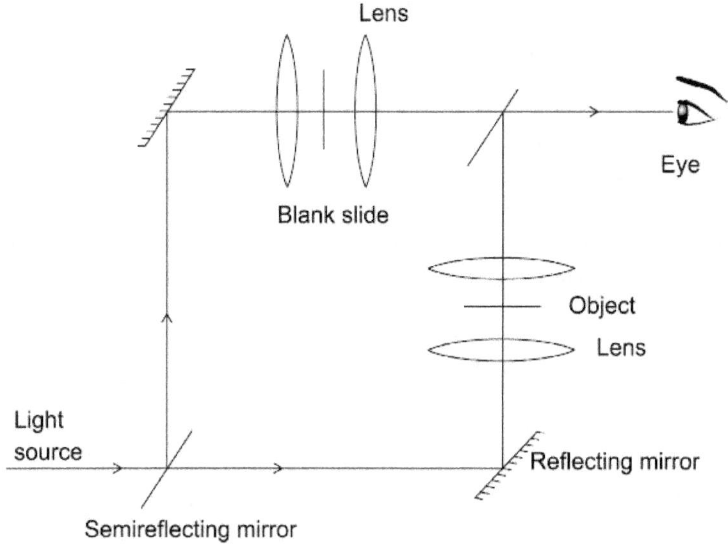

Figure 4.6. Optical principle of interference microscope.

surrounding medium (n_m). The phase change or retardation (T) can be determined using the following formula:

$$n_o - n_m = T/t$$

or

$$T = t\,(n_o - n_m)$$

The dry weight of the object is related to its refractive index so that the dry weight of the object can be determined by using the formula,

$$C_o = \frac{100\,(n_o - n_w)}{X}$$

where,

C_o = % of concentration of dry matter

n_o = Refractive index of the object

n_w = Refractive index of water

X = Constant.

Uses   Interference microscope is useful to measure the phase differences in cellular structures and their dry weight. In other words, simultaneous determination of the thickness of the object and the concentration of dry matter and water content. Therefore, the concentration of proteins, lipids and nucleic acids can be determined by using interference microscope.

4.5 POLARIZATION MICROSCOPE

The optical principle of polarization microscope is also the same as in the compound microscope but differs in having two polarizing optical devices (polarizer and analyser) as shown in the Figure 4.7.

The polarizer is formed by calcite sheet which is placed between the incident light and the specimen and another one is analyser which is placed above the objective lens. Instead of ordinary light, the polarizer transmits only plane polarized light which vibrates only in one direction. The analyser is rotatable and its rotation through 360^0 will cause alternate bright and dark field of vision for every 180° turn. In crossed position of polarizer and analyser, the polarized light is not transmitted and so the specimen is rotated to have maximum and minimum brightness. When the specimen is rotated at 45° to the polarizer and analyser, brightness is achieved.

Through an isotropic object, the polarized light travels with the same velocity and the refractive index will be the same in all directions. An anisotropic object is birefringent because the polarized light, when passed through it, becomes split up into two rays namely an ordinary ray and an extraordinary ray. These two rays are perpendicular to one another and have different velocities. If the velocity of the extraordinary ray is greater than that of the ordinary ray, then the birefringence is positive. If the velocity of the ordinary ray is greater than the extraordinary ray, then the birefringence is negative.

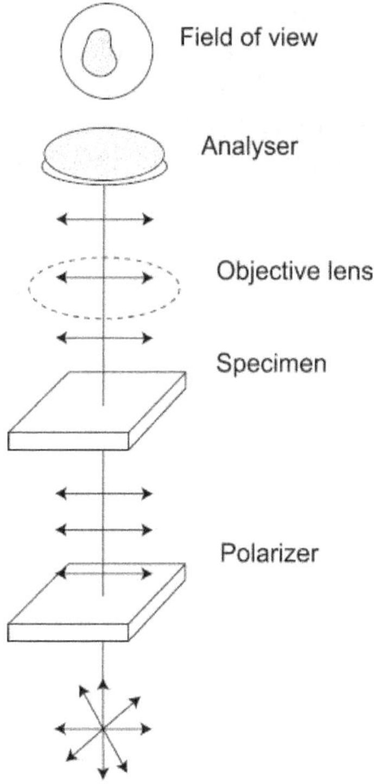

Figure 4.7. Optical principle of polarization microscope.

Uses Polarization microscope is used to study the structure of mitotic spindle and nerve fibres. Mostly, the biological fibres like proteins exhibit positive birefringence while the nucleic acids exhibit negative birefringence.

4.6 ULTRAVIOLET MICROSCOPE

The optical principle of ultraviolet microscope is more or less similar to that of a compound microscope except the following differences and its optical principle is schematically represented in the Figure 4.8. In this microscope, ultraviolet or infrared light is used as source of illumination instead of ordinary light. The ultraviolet light is split up into two beams one of which is passed through the microscope and another one to a photoelectric cell. When the ultraviolet beam passes through the object its intensity is decreased and this can be determined by comparing the photoelectric current of the two beams. The absorption of ultraviolet radiation by the object is recorded by the photoelectric cell present in the body of the microscope.

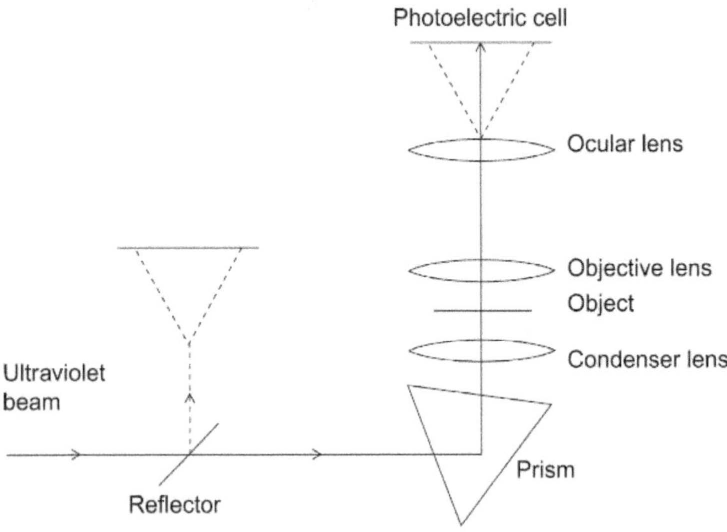

Figure 4.8. Schematic representation of ultraviolet microscope.

In this microscope, the lens systems, slides, cover glasses and other opticals are made of mineral fluorites or quartz. As ultraviolet rays cannot be perceived by human eye, the image of the object is examined with the help of a fluorescent screen or a photographic plate.

Uses   Unstained preparation of living cells can be observed using ultraviolet microscope as the nucleoproteins in living cells strongly absorb the ultraviolet rays. This helps to identify the quantity and distribution of nucleoproteins in cells. This microscope is also useful to study absorption of nucleic acids and proteins by the cell structures as well as the metabolic changes in cell division, growth and differentiation.

4.7 FLUORESCENT MICROSCOPE

The optic principle employed in fluorescent microscope is also similar to that of a compound microscope except for the incorporation of two filters to the optic system and is diagrammatically represented in Figure 4.9. Among the two filters, the source filter is placed between the source of light

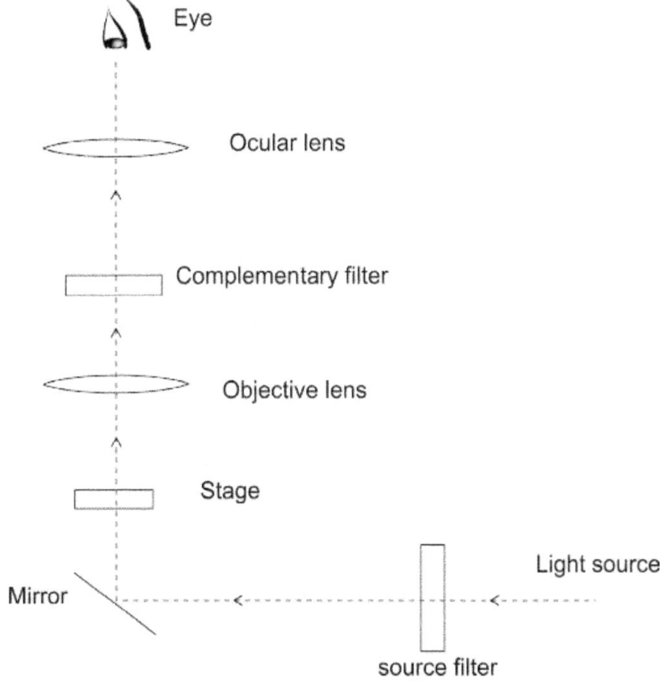

Figure 4.9. Optical principle of fluorescent microscope.

and mirror and the complementary filter in front of the ocular lens in the body tube. The former transmits the ultraviolet rays towards the object whereas the latter absorbs the rays, which are not absorbed by the object.

On illumination by ultraviolet radiation, many substances are capable of absorbing light of different wavelengths. This property is called fluorescence and the substances are called fluorescent substances. Biological molecules such as proteins and carbohydrates are treated with fluorescent dyes and examined.

Uses It aids in locating some specific components in cells and to localize the antigens for a number of viruses.

4.8 ULTRAMICROSCOPE (OR) DARK-FIELD MICROSCOPE

The optic principle of dark-field microscope is also the same as that of a compound microscope except for the nature of condenser lens which is shown in the Figure 4.10.

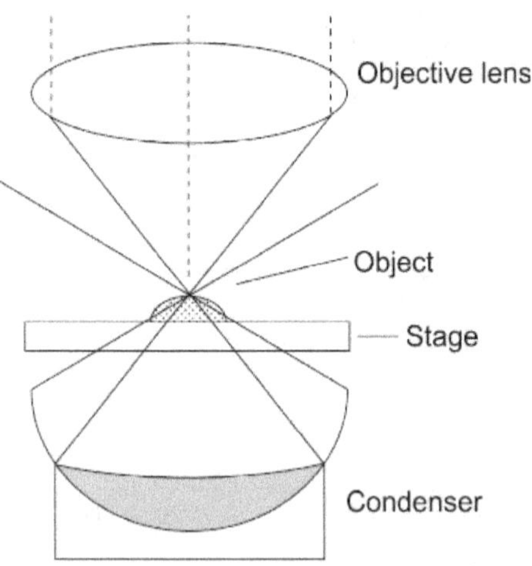

Figure 4.10. Optical principle of ultramicroscope.

The condenser is specially designed with a blackened area at the centre so that light is reflected to its outer edge. As a result, the condenser illuminates the object obliquely and the object looks bright in a dark background.

Uses   Ultramicroscope is useful in the study of living cells in which the cytoplasmic organelles are brightly illuminated whereas the cytoplasm remains dark. Thus, it finds great use in parasitology and bacteriology.

4.9 ELECTRON MICROSCOPE

Electron microscope uses high-speed electrons of short wavelength as source of illumination generated by high voltage and electromagnetic lenses along with photographic plate within a vacuum chamber. It is the only instrument available to study the biological ultrastructure and has a very high resolving power with a magnification power of 100000X or more. The electrons possess wave-like property like that of a beam of electromagnetic radiation. The energy of electrons is determined by their wavelength and their acceleration depends on the voltage.

Preparation of specimen for electron microscopy   For electron microscopic studies, the biological samples should be prepared with great care as the biological molecules disintegrate in vacuum. The biological structures (carbon, hydrogen, oxygen and nitrogen) possess low atomic number but electron microscopy requires objects of higher atomic number for greater resolution. Therefore the biological structures should be added with heavy atoms (fixation). The specimens can be fixed either by chemical fixation or by freezing. In chemical fixation, mostly osmium tetroxide, glutaraldehyde, potassium permanganate, formalin, etc. are used. These are high molecular weight chemical preservatives and form covalent bonds with biological molecules especially proteins and lipids.

In general, wet samples are dehydrated by increasing concentrations of ethanol or acetone followed by propylene oxide. The processed material is embedded in a hard embedding media such as araldite, vestopal-W, Epon-812, etc. Sections are cut by using ultramicrotome with a glass or diamond knife (50–100 mµ thickness). The extremely thin sections are mounted on a grid (a perforated metal disc) covered with parlodoan and can be examined directly. As electron microscope produces black and white photographic image of the object, there is no need for staining of sections. However, the use of positive staining provides a better contrast to the specimen. Therefore, solutions of heavy metals such as phosphotungstate, lead hydroxide, lead acetate, etc. are used to get contrast. The grid with the specimen is placed in between the condenser coil and the objective coil and the object is viewed on a photographic plate or on a fluorescent screen.

Uses It finds great use in the study of the ultrastructure of cells, to identify plant and animal viruses and to study the distribution of nucleic acids and enzymes within the cells. The following types of electron microscopes have been designed.

4.9.1 Transmission Electron Microscope (TEM)

Here, the electrons emitted by the cathode ray are passed through the specimen stained with metallic stains and focused on a photographic plate to get a black and white image of the specimen.

The optic principle of electron microscopy is the same as that of light microscopy. (Figure 4.11) except for the following differences:

i.   The source of illumination is high velocity electron beam emitted by cathode filament.

ii. Instead of optical lenses, electromagnetic lenses are used.

iii. As the atoms in air absorb the electrons and travel considerable distances only in vacuum, the entire setup is kept in a vacuum-tight tube.

iv. The images of the object are seen only on photographic plate.

Moreover, examination under TEM requires extremely thin sections as the electrons have low penetration power. Therefore, the sections are cut to 50–100 nm in thickness. As a result, the sections are to be supported by very thin film of carbon or other substances and should be stained with heavy metals such as osmium or gold. The sections are thoroughly dried in vacuum, so that live cells cannot be studied using electron microscope.

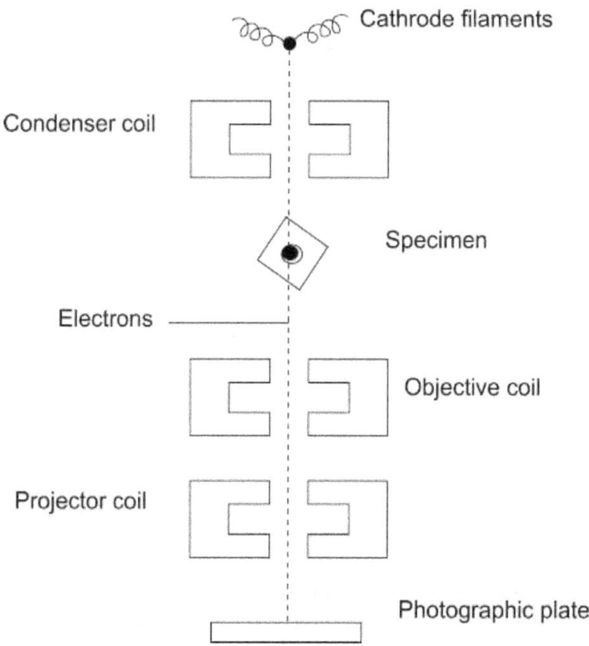

Figure 4.11 Optical principle of TEM.

4.9.2 Scanning Electron Microscope (SEM)

Diagrammatic structure of scanning electron microscope is shown in Figure 4.12. Here the reflected electrons (secondary electrons) from the specimen are collected by a grid or collector which produces a flash of light on the scintillator. The light from the scintillator is amplified by a photomultiplier and scanned. The scanned image can be viewed on a screen or can be photographed.

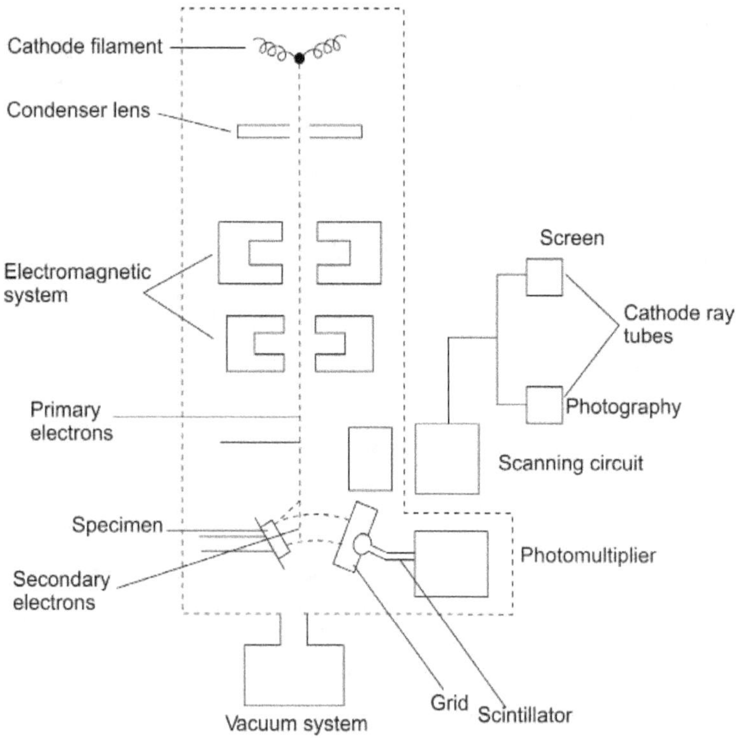

Figure 4.12. Diagrammatic structure of SEM.

Uses　At present SEM is more popular and is useful for the direct observation of the sections to get images of the surface of the specimen.

4.9.3 Electron Cryomicroscopy (EC)

This microscope is nothing but a modified design of TEM and is useful to observe unfixed, unstained and hydrated biological specimens. The sample in aqueous suspension is placed on a grid in the form of a very thin film and is frozen in liquid nitrogen. Water is not evaporated even in vacuum at a very low temperature of about $-196°C$, therefore the frozen sample in aqueous suspension can be examined without fixation under a cryoelectron microscope. The structure of the object can be photographed using low-dose electrons. Computer programmes can analyse three-dimensional structure of molecules so that molecular models of the specimen are made.

Uses   Larger proteins, which are difficult to crystallize, can be studied by cryoelectron microscope. Here the protein sample is preserved by rapid freezing with liquid helium, which is then hydrated and examined to get molecular models of proteins.

4.9.4 Scanning Tunneling Electron Microscope (STEM)

The basic principle of STEM involves the probing of biological molecules through the movement of electrons from one region to another requiring quantum mechanical tunnelling. As the quantum mechanical wave has the ability to pass though an energy barrier, the barrier is made to possess near-atomic dimensions in order to measure the tunnelling current.

The STEM as shown in Figure 4.13 consists of a sharp, electrically conducting tip which produces a tunneling current between the tip and the specimen. An empty space acts as the barrier in which two electrical conductors are separately placed at a distance. Therefore, the unwanted atoms can also enter into this space. This facilitates an easy observation of the biological surfaces even when they are surrounded by water. To achieve an atomic resolution, piezoelectric effect is arranged with suitable crystals (Figure 4.14).

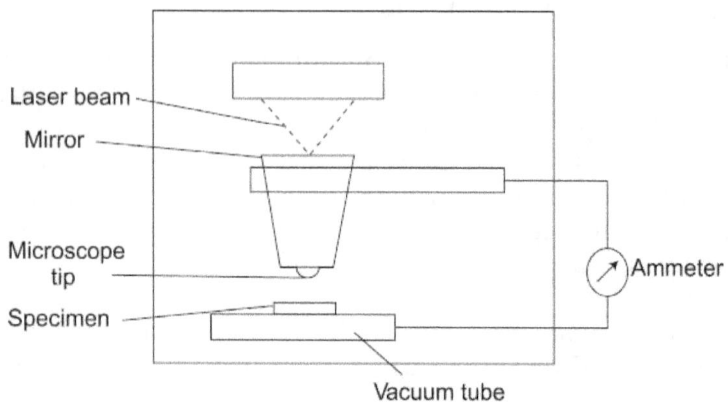

Figure 4.13. Diagrammatic structure of STEM.

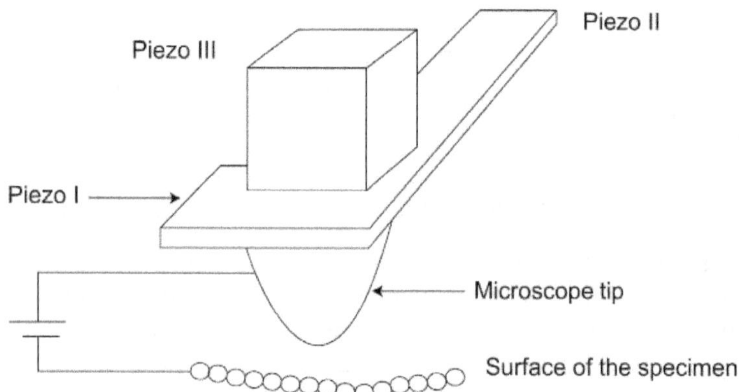

Figure 4.14. Tip of STEM probing the surface of the specimen (Diagrammatic).

The physical dimension of the crystal can be changed by varying the applied voltage so as to control the position of the tip of the microscope to lie in the plane of the specimen surface. Thus the sharp tip of the microscope is brought very close to the surface of the specimen at a particular distance. The electrons from the surface of the specimen tunnel across the barrier space producing a tunneling current, which is recorded.

The recording of the vertical portion of the tip at each position gives the structure of the surface of the specimen.

Uses   STEM is mainly used to study the samples of macromolecules such as proteins, nucleic acids and other macromolecular assemblies.

PHOTOMETRY OR ABSORPTIOMETRY

Photometry is a technique used to determine the concentration of substances by using the property of absorption of light of definite wavelength by the molecules, especially biochemical compounds. This method works on the principle of **absorption laws** given by Beer and Lambert.

Beer's law   when a monochromatic light ray passes through an absorbing medium, its intensity decreases exponentially as the concentration of the absorbing medium increases. (Figure 4.15).

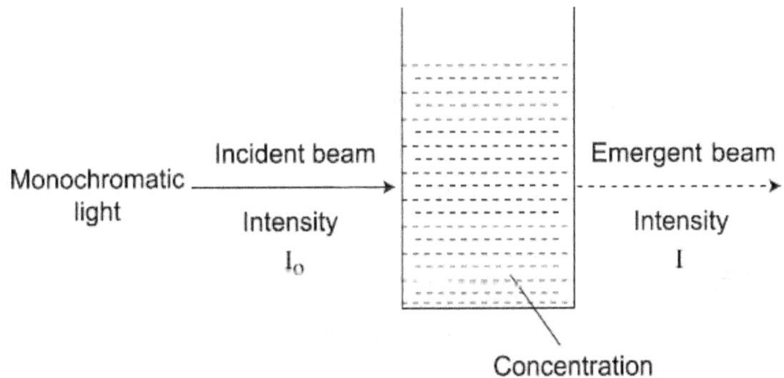

Figure 4.15. Diagram representing Beer's law.

i.e.
$$\log \frac{I_o}{I} = K_1 C$$

where,

I_0 = intensity of incident beam

I = intensity of emergent beam

C = concentration

K_1 = absorption coefficient

Lambert's law   When a monochromatic light ray passes through an absorbing medium, its intensity decreases exponentially as the length of the absorbing medium increases (Figure 4.16).

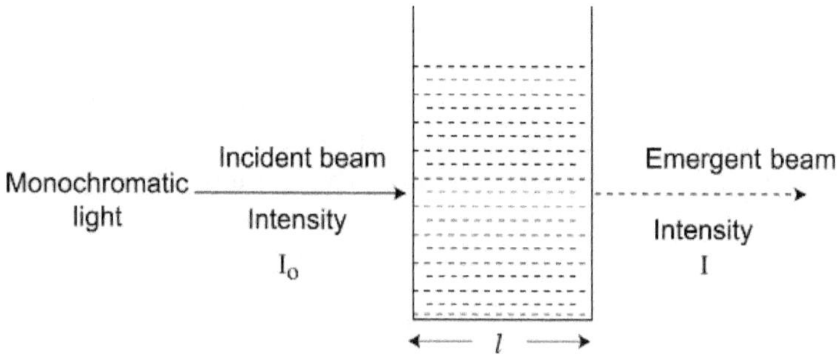

Figure 4.16. Diagram representing Lambert's law.

i.e.
$$\log \frac{I_o}{I} = K_2\, l$$

where,

l = thickness of the absorbing medium

K_2 = absorption coefficient

Beer–Lambert's law   The above laws can be combined and mathematically expressed in the following manner:

$$\log_{10} \frac{I_o}{I} = \log_{10} KCl$$

Optical density (OD) or **extinction (E)** or **absorbance** is the logarithmic ratio of I_o and I and so the optical density can be defined as the logarithm of the ratio between the intensities of light beam before entering and after leaving the sample. On substitution of E in the above equation,

$$E = KCl$$

That is, when light passes through a solution, the extinction is directly proportional to the length of the solution (l) and the concentration of the absorbing substance (C).

Transmittance (T) is the ratio of intensity of emergent light to that of incident light. Therefore transmittance is inversely proportional to optical density.

i.e.
$$T = \frac{I}{I_o}$$

The percent transmittance is the transmission of light when $I_o = 100$. Now extinction and percent transmittance can be related.

$$E = KCl$$

$$\text{or } \log_{10} \frac{I_o}{I}$$

where,

$$I_o = 100$$

$$E = \log_{10} \frac{100}{I}$$

or
$$E = \log_{10} 100 - \log_{10} I$$

i.e.
$$E = 2 - \log_{10} I$$

The above principle is employed in all instruments of photometry.

4.10 IMPORTANT COMPONENTS IN INSTRUMENTS OF PHOTOMETRY

All instruments of photometry possess the following components:

4.10.1 Light Source

Tungsten filament lamp is most commonly used and it produces light over the whole of the visible spectrum and in near infrared and near ultraviolet regions. Biological substances are mostly colourless and do not absorb in the visible region but absorb in ultraviolet region of the spectrum. For these samples, **hydrogen-discharge lamp** made of quartz is used. A **deuterium discharge lamp** is used when high energy levels are required in the ultraviolet region.

4.10.2 Monochromator

The continuous spectrum produced by the lamp contains many wavelengths which are not completely absorbed by the sample. Therefore, it is necessary to select a particular wavelength by some arrangement. This wavelength selector is called **monochromator,** which may be a simple filter, prism or diffraction grating. The filters absorb or reflect light except one band of wavelength. The gelatin filters are made of gelatin, coloured with organic dyes and sealed between glass plates. Tinted glass filters are coloured glass filters mostly used in modern instruments. Interference filters are improved glass filters in which a layer of transparent material like calcium or magnesium fluoride is coated on each side with a thin layer of silver placed between two glass plates.

The colour of the filter is complementary to the colour of the sample solution as shown in Table 4.1.

A prism disperses the polychromatic light into a spectrum. Glass prisms are used in visible spectrum but prisms made of

Table 4.1. Colour of the filter complementary to the colour of the solution.

Colour of the solution	Filter colour
Red-orange	Blue-blue green
Blue	Red
Green	Red
Purple	Green
Yellow	Violet

quartz or fused silica are used in ultraviolet light. But quartz prisms exhibit circular double refraction. This problem is overcome by joining two pieces of opposite-handed quartz crystals to get a prism as in the Figure 4.17.

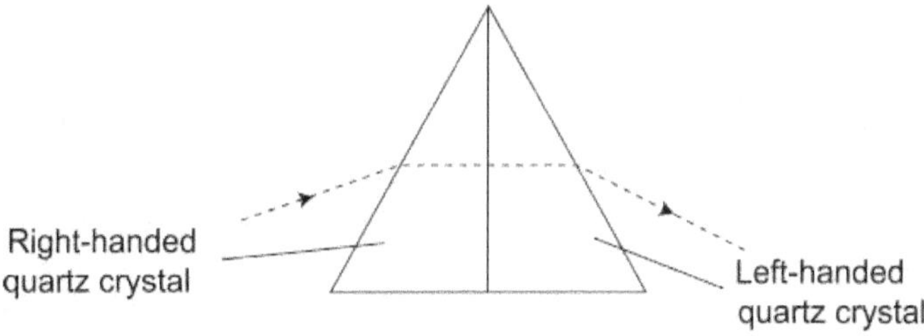

Figure 4.17. Quartz prism with right- and left-handed crystals.

In this prism, the circular double refraction of right-handed quartz is balanced by an equal and opposite effect in the left-handed quartz. A 30° prism made of single type of quartz crystal with a mirror arrangement on the back face of the prism is also used (littrow mounting). A prism monochromator is not used alone but with a series of lenses and slits. The diffraction gratings are the best monochromators to disperse the light rays and are of two types. A **transmission diffraction grating** is a transparent plate on which a large number of parallel lines are ruled. A **reflection diffraction grating** is one in which a large number of grooves are ruled on to a highly polished reflecting surface.

4.10.3 Sample Holders

The selection of sample holders depends on the type of the instrument. In simple colorimeters, glass test tubes with walls having uniform thickness are used. In spectrophotometers, rectangular glass cells or cuvettes are used. Each cuvette has ground glass on two sides and the other two sides are polished to be optically flat. Quartz cells are used in ultraviolet region of the spectrum. The inner optical path length of standard cells is about 1 cm. To measure small quantity of samples, 1-mm cells are available. Weakly absorbing samples are measured by using cells of longer path lengths up to 10 cm.

4.10.4 Light Sensitive Detectors

The detectors collect the electrons liberated by the incident light from the metal and measure the electrons as current, which is proportional to the light intensity. In modern photometers, two types of detectors namely **photocells** (photoelectric cells or photovoltaic cells) and **photomultiplier** tubes are used. A photocell uses crystalline semiconductor materials such as cadmium sulphide or silicon but mostly selenium. The structure of a selenium photocell is represented in the Figure 4.18. It consists of a steel base plate, which acts as anode on which molten selenium in the form of a thin coating is applied and covered by a thin transparent film of silver. Electrons travel easily from the selenium to silver layer, which acts as a collecting electrode (cathode) for electrons liberated from the selenium layer by incident light. A micrometer measures the current. The photocell can be used for the entire visible spectrum and its output need not be amplified. A photomultiplier tube is used to amplify the current and its structure is shown in the Figure 4.19. Here the cathode and anode are sealed within an evacuated glass tube with additional electrodes called dynodes between them. When light passes onto the cathode, the liberated electrons are accelerated towards the first dynode and then towards the next dynode until the last one. In this way the current becomes amplified when the electrons reach the collecting anode. Many spectrophotometers use two photomultiplier tubes to cover the visible spectrum (one is sensitive to red and the other one to blue).

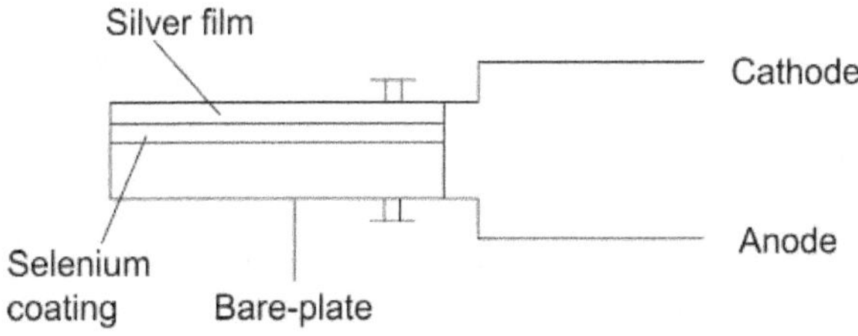

Figure 4.18. A selenium photocell.

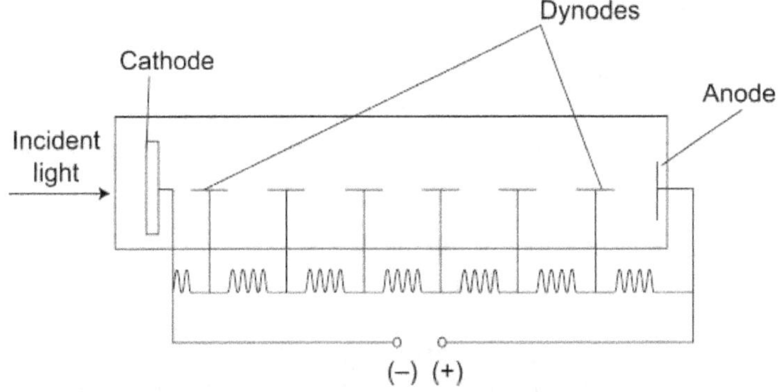

Figure 4.19. A photomultiplier tube.

4.11 COLORIMETER

Colorimeter is useful to determine the concentration of coloured samples by measuring the amount of light intensity actually absorbed by the sample. The components of a colorimeter are shown in Figure 4.20.

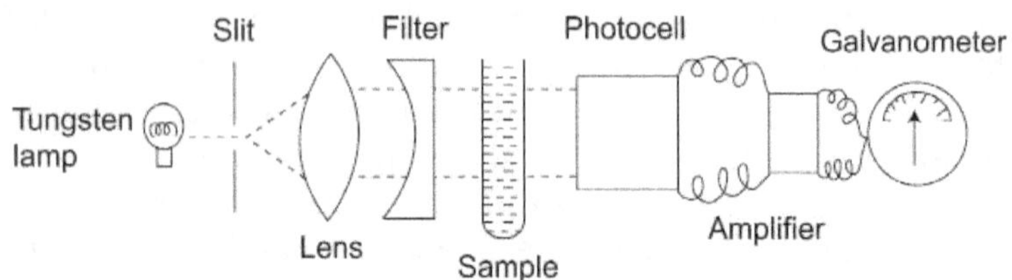

Figure. 4.20. Components of a colorimeter (diagrammatic).

A tungsten lamp provides white light as a steady source of radiation through a slit. A condenser lens makes the light beam parallel and to fall on the solution. A coloured glass filter is used as monochromator, the colour of which is complementary to the colour of the sample. A photocell produces electrical current in proportion to the intensity of light falling on it. The current produced by the photocell is amplified by an amplifier. A galvanometer calibrated with optical density and percent transmittance in log scales gives absorbance readings directly.

A standard graph is plotted by using known concentrations of the substance relating the concentration and optical density. From this standard graph, the concentration of unknown solution or sample can be arrived at.

SPECTROPHOTOMETER

Spectrophotometer is nothing but the extension and refinement of colorimetry. By using spectrophotometer, the concentration of colourless substances can be determined or the substances may be converted into coloured compounds and then the concentration is estimated. By this instrument, any material that absorbs visible light, ultraviolet light or infrared light could be estimated.

The optic principle of spectrophotometer is very similar to that of a colorimeter and its components are shown in Figure 4.21. A steady source of radiation is produced by a tungsten lamp and a light surface is provided to get light rays of required wavelength. A prism or a grating made of crystal, quartz or fused silica produces a monochromatic spectrum. An adjustable slit allows only a particular portion of the spectrum to pass through the detector. The amplifier and recorder perform functions as in colorimeter.

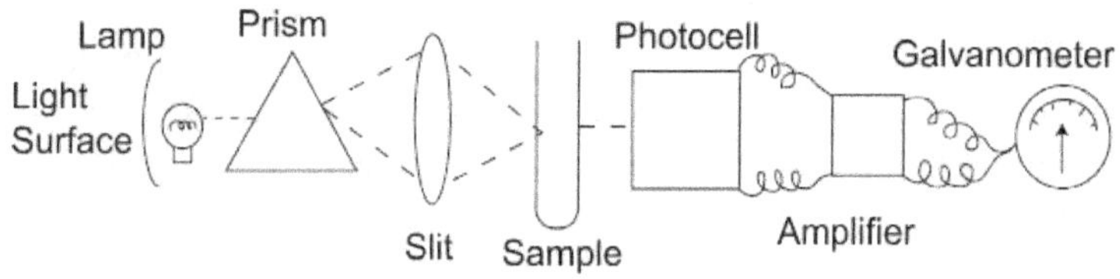

Figure 4.21. Components of a spectrophotometer (diagrammatic).

4.12 ULTRAVIOLET AND INFRARED SPECTROPHOTOMETERS

In both instruments, the components are the same as in spectrophotometer except that the source of illumination is a special deuterium lamp. In addition, quartz or fused silica cuvettes are used as sample holders. The ultraviolet spectrophotometer works at wavelengths below 400 mµ and the infrared spectrophotometer above 800 mm. Many spectrophotometers are also designed to work over a wide range of wavelengths ranging from 185–3500 mµ.

Thus the spectrophotometer is an improved instrument over the colorimeter as it works at various ranges of wavelengths. It is useful for the determination of concentration of samples, identification of materials and estimation of rate of reactions. Ultraviolet and infrared spectrophotometers are useful in the estimation of biological compounds like proteins, vitamins, hormones and nucleic acids as well many dehydrogenase enzymes.

4.13 FLAME PHOTOMETER

The optical principle and detector systems of flame photometry are very similar to that of spectrophotometry except that the function of the sample cell is substituted by the flame. When a metallic salt is introduced into the non-luminous flame, it ionizes and emits light with characteristic wavelength. The correlation of the emission intensity of an element with its concentration is the basis of flame photometry, which includes **emission flame photometry**, and **atomic absorption photometry**. The flame photometer consists of the following components:

Flame   It is important for the atomization of the sample. When an aqueous sample is introduced into a flame as a mist of fine droplets, water is quickly evaporated leaving the dry salt, which melts, vaporizes and decomposes into free atoms. Some elements such as alkali metals require only the thermal energy of the flame (low temperature) to excite the atoms. Some metals like aluminium require a high temperature. In emission flame photometer, a burner similar to bunsen burner produces low temperature flame. But the fuel gas reaches the face of the burner through a number of openings so that a circular-shaped flame is produced. In atomic absorption flame photometer, a slot-type burner is used so that a ribbon-shaped flame is produced.

Nebulizer   Before introducing into the flame, the sample must be converted into a mist of fine droplets. This process is called as **nebulization**, which is achieved by a **nebulizer.** The nebulizer consists of a capillary inlet opening into a spray chamber. When the sample is allowed to pass into the capillary tube, it emerges from the tip of the capillary and is broken down into fine droplets of varying sizes by flowing the compressed air. The spray chamber has a series of mechanical baffles, which drain the larger droplets. Therefore most of the sample introduced into the spray chamber is lost. To overcome this problem, an impact bead made of some chemically inert material like fused silica, is placed in the spray chamber at a distance from the nebulizer tip as in the Figure 4.22. The impact bead breaks up the large droplets, thus reducing their size. In this way, the efficiency of the nebulizer is improved.

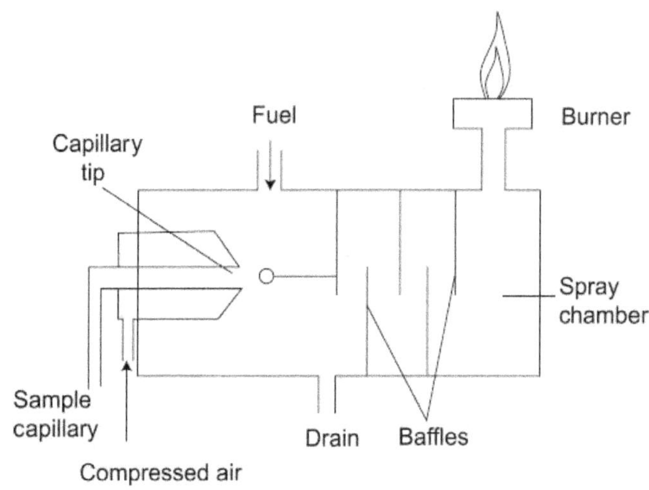

Figure 4.22. An improved nebulizer and burner.

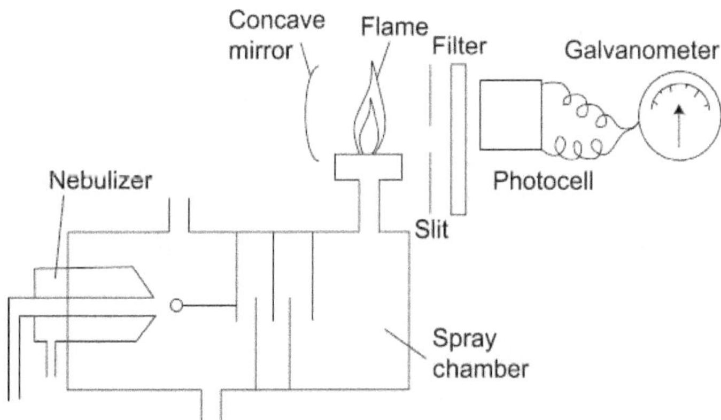

Figure 4.23. Components of emission flame photometer.

Optical systems and detectors The optical principle and the detectors in flame photometer are very similar to that of the spectrophotometer as shown in the Figure 4.23. The wavelength of light is selected by filter, prism or diffraction grating and the detector is a photocell or a photomultiplier tube.

Uses The flame photometer finds great use in determining alkali metals mainly sodium and potassium as well as lithium and calcium.

4.14 ATOMIC ABSORPTION FLAME PHOTOMETER

The atomic absorption flame photometer is very similar to that the emission flame photometer except for the provision of a source of radiation, which is made to traverse the flame. The atoms, which are excited to emit light in a flame, are capable of absorbing radiations of their resonance wavelength. A particular resonance wavelength from a source of light cannot be isolated with the help of a prism or a diffraction grating. Using hollow-cathode discharge lamps, which produce monochromatic radiation characteristic of the element to be analysed, solves this problem. The structure of a cathode discharge lamp is shown in the Figure 4.24. It consists of two electrodes of which the cathode is a hollow tube coated with the element to be studied. The electrodes are sealed in glass envelope, which is filled with an inert gas like argon or neon. On application of high voltage to the lamp, electrons from cathode travel towards the anode. The travelling electrons collide with atoms of inert gas and ionize them into large positive ions, which move towards the cathode. Now the gas ions and electrons collide with metal atoms in the cathode. As a result, the excited metal atoms accumulate around the cathode and emit radiation, which comes out as a beam containing complete emission spectrum through the end window of the lamp. Chopping or modulating the emergent beam with the help of a rotating shutter or an electronic device overcomes the problem of background radiation due to other elements of the sample and radiation from the flame.

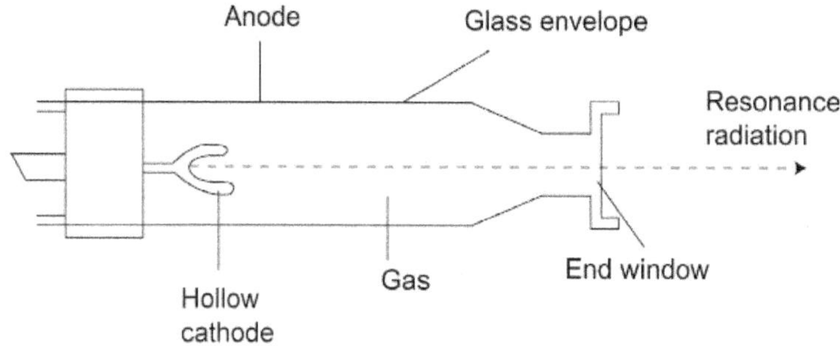

Figure 4.24. A hollow–cathode discharge lamp.

Diagrammatic structure of an atomic absorption flame photometer is given in the Figure 4.25. The burner is supplied with a specific gaseous mixture to produce characteristic type of flame for different elements as in Table 4.2.

Table 4.2. Types of flames for various elements.

Flames	Elements to be studied
Air-acetylene	Tin, molybdenum, chromium, copper, iron, potassium and magnesium.
Nitrous oxide–acetylene	Magnesium, calcium, strontium, barium, tin, chromium, molybdenum and boron.
Air–hydrogen	Arsenic, selenium and tin
Nitrogen–hydrogen–entrained air	Arsenic, selenium, cadmium, mercury, tin, lead, calcium and zinc.
Arogon–hydrogen–entrained air	Sodium and potassium
Nitrous oxide–propane	Magnesium and calcium

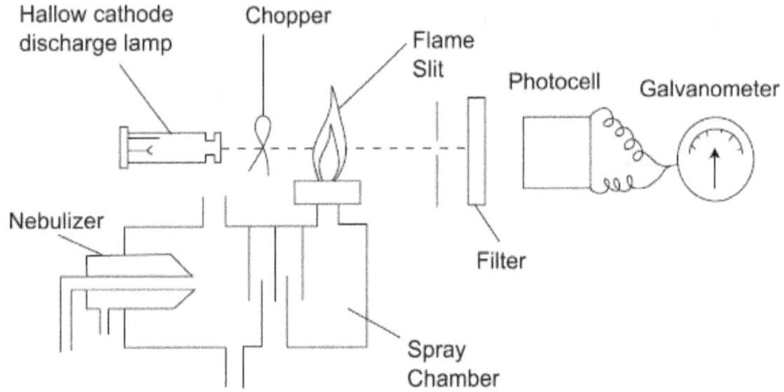

Figure 4.25. Components of atomic absorption flame photometer.

Uses   This instrument is useful to determine a wide range of elements.

REVIEW QUESTIONS

1. Explain the process of dispersion and diffraction of light.

2. Define magnification and resolving power.

3. Describe the optical principle of a compound microscope.

4. What is the significance of oil immersion objective?

5. Describe the optical principle of phase contrast microscope.

6. Describe the optical principle of interference microscope.

7. Describe the optical principle of polarization microscope.

8. Describe the optical principle of ultramicroscope.

9. Describe the optical principle of fluorescent microscope.

10. Describe the optical principle of transmission electron microscope (TEM).

11. Describe the optical principle of scanning electron microscope (SEM).

12. Describe the optical principle of scanning tunnelling electron microscope (STEM).

13. Describe the optical principle of electron cryomicroscope.

14. What are the steps involved in preparing a specimen for electron microscopy.

5

BIOPHYSICAL PHENOMENA IN BIOCHEMICAL STUDIES

5.1 HYDROGEN ION CONCENTRATION (pH)

In living organisms, all biochemical reactions occur in an aqueous medium. The substances in the cytosol, blood, body fluids, glandular secretions, etc. include free ions, molecules and macromolecules, which carry ionizable groups. These ions determine the biochemical and biophysical changes in living systems.

They are electrically charged particles with either positive or negative charge. The important cations are H^+, Ca^{++}, Na^+, K^+, Mg^{++}, etc. and anions are OH^-, HCO_3^-, Cl^-, HPO_4^-, $H_2PO_4^-$, etc. The control mechanisms in biological processes are sensitive to the concentration of H^+ and OH^- ions in the medium because the acidity or alkalinity of a solution is determined by its H^+ ion concentration or pH (Potential of Hydrogen).

An excess of H^+ ions over OH^- ions is considered as acidity and an excess of OH^- ions over H^+ ions as alkalinity. In other words, the intensity of acidity depends on the amount of H^+ ions in excess and that of alkalinity depends on the excess of OH^- ions. In a solution, if the quantity of OH^- ions is the same, then it is neutral. For example, pure water contains H^+ and OH^- ions in equal proportions. An acid solution is one in which $[H^+]$ is greater than 10^{-7} and an alkaline solution is one in which $[H^+]$ is lesser than 10^{-7}.

pH can be defined as the $-\log$ of the H^+ ion concentration. Therefore, the pH scale is logarithmic and is applicable only to solutions at normal temperature.

$$pH = -\log_{10}[H^+] \text{ or } \log_{10}\frac{1}{H}$$

For example, pure water has a pH of 7.

i.e.
$$pH = -\log_{10} 10^{-7}$$
$$= -(-7) = 7$$

For every molecule, the degree of dissociation is described by an equilibrium or dissociation constant. For example, the dissociation constant for H_2O is as follows:

$$H_2O \rightleftharpoons H^+ + OH^-$$

$$\therefore K = \frac{[H^+][OH^-]}{[H_2O]}$$

In water and dilute aqueous solutions, the concentrations of H^+ and OH^- ions remain constant so that

$$K_w = 1.00 \times 10^{-14}$$

Therefore, both H^+ and OH^- ions in pure water have the concentration of 10^{-7} moles/litre whereas in acidic and alkaline solutions, the concentrations are 1.00×10^{-14}. A neutral solution has pH 7 and ionizes as follows:

$$[H^+] \times [OH^-] = 1 \times 10^{-14}$$

A solution with pH less than 7 is acid and higher than 7 is alkaline. That is why, the pH scale runs from 0–14 as shown below.

5.1.1 pH Scale

H^+ ion	1	10^{-1}	10^{-2}	10^{-3}	10^{-4}	10^{-5}	10^{-6}
	0	1	2	3	4	5	6

Acidic

10^{-7}	10^{-8}	10^{-9}	10^{-10}	10^{-11}	10^{-12}	10^{-13}	10^{-14}
Neutral	8	9	10	11	12	13	14

Alkaline

In any aqueous solution, the concentration of H^+ and OH^- ions will be only 10^{-14}. That is, the highest pH will be only 14 (Table 5.1).

Table 5.1. Concentrations of H^+ and OH^- ions on pH scale.

Concentration of H^+ ions	Concentration of OH^- ions	pH
10^{-1}	10^{-13}	1
10^{-3}	10^{-11}	3
10^{-6}	10^{-8}	6
10^{-10}	10^{-14}	10
10^{-12}	10^{-2}	12
10^{-14}	10^{-0}	14

The acids are proton donors and bases are proton acceptors. Strong acids and bases dissociate completely in dilute aqueous solutions into protons and OH^- ions respectively. Therefore, their dissociation constants will be infinity. This is because the pH of a solution mainly depends on the activity of ions present in it. In more concentrated solutions, the ions begin to interact with each other as well as with solvent molecules, so that the activity of ions is lesser at high concentrations. But in biological structures, most of the acidic and basic substances are weak acids and weak bases and dissociate only partially. In an aqueous solution of a weak acid, there is equilibrium between the acid and its conjugate base which can accept a proton to form an acid again. The conjugate bases may not contain OH^- groups but they increase the concentration of OH^- ions by getting a proton from water. A few weak acids and their conjugate bases are given in Table 5.2.

Table 5.2. Weak acids and their conjugate bases.

Acid (proton donor)	Conjugate base (proton acceptor)
Acetic acid (CH_3COOH)	Acetate ion (CH_3COO^-)
Carbonic acid (H_2CO_3)	Bicarbonate ion (HCO_3^-)
Bicorbonate (HCO_3)	Carbonate ion (CO_3^-)
Formic acid (HCOOH)	Formate ion ($HCOO^-$)
Phosphoric acid (H_3PO_4)	Dihydrogen phosphate ion ($H_2PO_4^-$)
Dihydrogen phosphate ($H_2PO_4^-$)	Monohydrogen phosphate ion (HPO_4^-)
Monohydrogen phosphate (HPO_4^-)	Phosphate ion (PO_4^-)

In living organisms, H^+ ions originate form water, which is a covalent compound in which H and O atoms are linked together by covalent bonds. Though water is a neutral molecule, it acts as a very weak acid as well as a very weak base and therefore it has the lesser ability of ionization. This is because one H_2O molecule can transfer a proton to another resulting in the formation of hydronium ion (H_3O^+) and a hydroxyl ion (OH^-)

$$H_2O + H_2 \rightleftharpoons H_3O^+ + OH^-$$

Thus, water is both a proton donor and a proton acceptor. Therefore, it is an amphoteric substance, which behaves both as an acid and base as shown below.

$$HOH \rightleftharpoons H^+ + OH^- \text{ (acid)}$$

$$HOH + H^+ \rightleftharpoons H_3O^+ \text{ (base)}$$

5.1.2 Determination of pH

The pH of a solution can be determined using any one of the following methods.

1. Henderson–Hasselbalch equation   Let us consider HA as a weak acid which ionizes in the following manner:

$$HA \rightleftharpoons [H^+] + [A^-]$$

$$\therefore \text{Dissociation constant (Ka)} = \frac{[H^+]+[A^-]}{[HA]}$$

$$H^+ = \frac{Ka \times [HA]}{[A^-]}$$

$$\log H^+ = \log Ka \times \log \frac{[HA]}{[A^-]}$$

$$-\log H^+ = -\log Ka - \log \frac{[HA]}{[A^-]}$$

$$\therefore pH = pKa + \log \frac{[A^-]}{[HA]}$$

$$= Pka + \log \frac{[salt]}{[acid]}$$

Thus the pH of any solution can be calculated by using Henderson–Hasselbalch equation, if its dissociation constant and concentration are known.

2. Indicators   pH indicators are certain dyes which determine H^+ ion concentration in a solution. These are the substances which exhibit characteristic colour with change in pH (Table 5.3). These behave as weak acids and dissociate in the following manner:

$$HI_n \rightleftharpoons H^+ + I_n$$

The dissociation constant of indicators can be shown by the equation given below:

$$KI_n = \frac{[H^+][I_n]}{HI_n}$$

In solutions having high concentrations of H^+ ions, the indicators do not completely dissociate. When the pH of the solution is raised, they dissociate completely and produce colour variations depending on the dissociation.

For instant determination of pH of the samples, indicators are used and compared with colour standards. If they occur in powder forms, then they are made into 0.1% aqueous solutions. pH papers impregnated with a mixture of indicators are also commercially available.

Table 5.3. Common indicators in biological studies.

Indicator	pH range	Color change
Alizarin yellow	10.0– 2.1	Yellow $\rightarrow$ brown
Bromophenol blue	3.0–4.6	Yellow $\rightarrow$ blue
Congo red	3.0–5.0	Blue $\rightarrow$ red
Litmus	3.1–4.4	Red $\rightarrow$ blue
Methyl orange	4.4– 6.6	Red $\rightarrow$ yellow
Methyl red	4.2–6.3	Red $\rightarrow$ yellow
Phenolphthalein	8.3–10.0	Colorless $\rightarrow$ red
Phenol red	6.8–8.4	Yellow $\rightarrow$ red
Thymol blue	1.2–2.8	Red $\rightarrow$ yellow
Topfer's reagent	2.9–4.0	Red $\rightarrow$ yellow

However, the results obtained by using indicators are only approximate and high levels of accuracy cannot be got in pH determination.

3. pH meter   pH meter is an electronic instrument which is used to determine pH and it gives constant and reproducible results. As a result, the disadvantages of human errors by the use of pH indicators and titrations can be eliminated.

5.2 pH METER

pH of sample solutions can be measured using coloured indicators which give only approximate results with visual judgment. Though titrations give good results, they are tedious and close results are not obtained by two titrations. All these defects are eliminated in the determination of pH of the samples by using an electronic instrument namely pH meter, which gives constant and reproducible results.

5.2.1 Principle

When a piece of metal is dipped into water, the atoms of the metal (+ve ions) diffuse into water and the metal becomes positively charged. In other words, the metal dissolves in water to some extent and equilibrium will be developed when the tendency of atoms to leave the metal is balanced by the tendency of the metal atoms in water to reattach to the metal. As a result, a potential is set up across the metal–water interphase. When the metal is dipped in an aqueous solution of one of its salts, the potential developed will depend on the concentration of the solution (Figure 5.1). Thus the concentration affects the tendency of metal atoms to dissolve in the solution. Here, the equilibrium will depend on two factors namely the solution pressure of the metal (P) and the osmotic pressure of ions in the solution (p).

When $P > p$, the metal produces positive ions till the equilibrium is obtained so that the metal becomes negatively charged. If $P > p$, the positive ions in the solution get deposited on the metal so that the metal becomes positively charged. On the other hand, if $P = p$, there is no ionization of metal ions without potential difference.

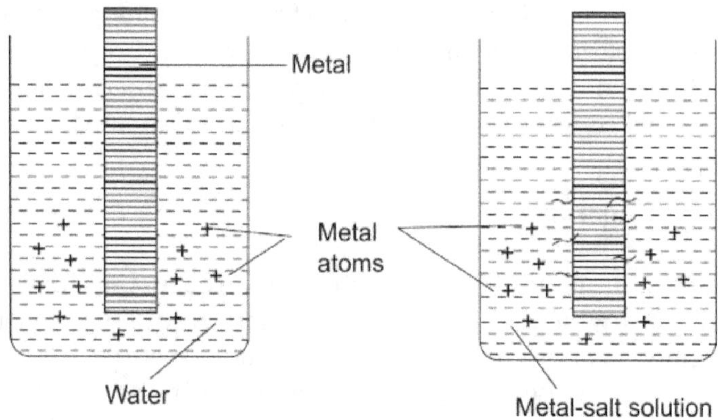

Figure 5.1. Dissociation of metal in water and metal–salt solution.

In this way, a piece of metal dipped into a solution will produce a potential. If electric current is used to measure this potential, the potential will be changed. Therefore, an electrometer, which draws very little current, must be used. Moreover, a second metal electrode in solution is to be provided to complete the electric circuit. When two electrodes of the same metal dipped into the solution of different concentrations are used, the potentials in the electrodes will be different. On electric supply, the current will pass from one electrode to another and each electrode will produce an electrode potential of its own. Now, the electrometer measures the differences in potential of the two electrodes. Thus, the ionic concentration of one cell can be determined by knowing the ionic concentration of the other cell.

5.2.2 Electrode System in pH Meter

The electrode system of a pH meter consists of a reference electrode and a glass or measuring electrode. The reference electrode when dipped in any solution develops a constant potential irrespective of the pH of the solution. On the other hand, the glass electrode is sensitive to the pH of the solution, therefore the pH measurement mainly depends on the glass electrode.

Reference electrodes

Hydrogen electrode   This electrode works on the principle that an inert metal like platinum readily absorbs hydrogen gas and produces an electrode potential on immersion in a solution containing H^+ ions.

It consists of a piece of platinum coil suspended in 1.18 M hydrochloric acid solution. Hydrogen gas is bubbled over the platinum which simply acts as an electrical conductor so that its potential is nearly zero (Figure 5.2). It is a inconvenient to use hydrogen electrode for everyday work because maintenance of hydrogen supply over the electrode could cause problems during measurement.

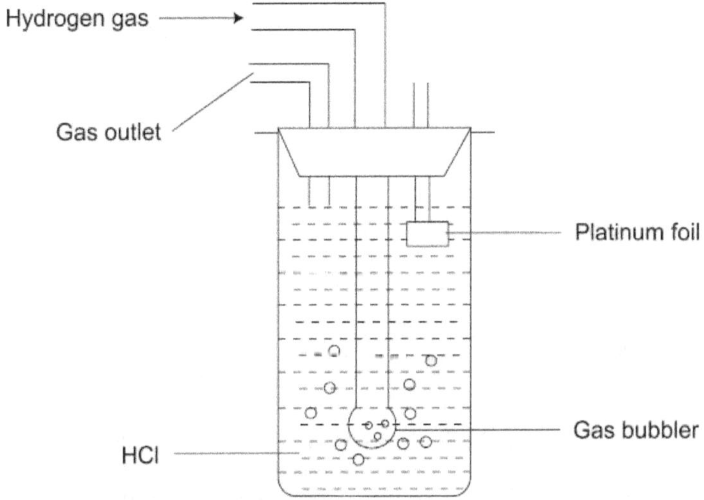

Figure 5.2. Structure of a hydrogen electrode.

Calomel reference electrode   This electrode works on the principle that the electric potential is generated in the sample solution when calomel is placed in potassium chloride solution.

It consists of a piece of platinum wire dipped into mercury in a sealed glass tube. A paste of calomel (mercurous chloride) is placed below the mercury layer with the help of cotton wool or sintered glass plug. The whole tube is filled with saturated KCl solution. A porous ceramic plug enables the contact of the electrode with outside solution. Through an opening with a rubber sleeve at the top, KCl solution is poured into the tube. Some KCl crystals should always be present at the bottom of the tube to ensure saturation (Figure 5.3).

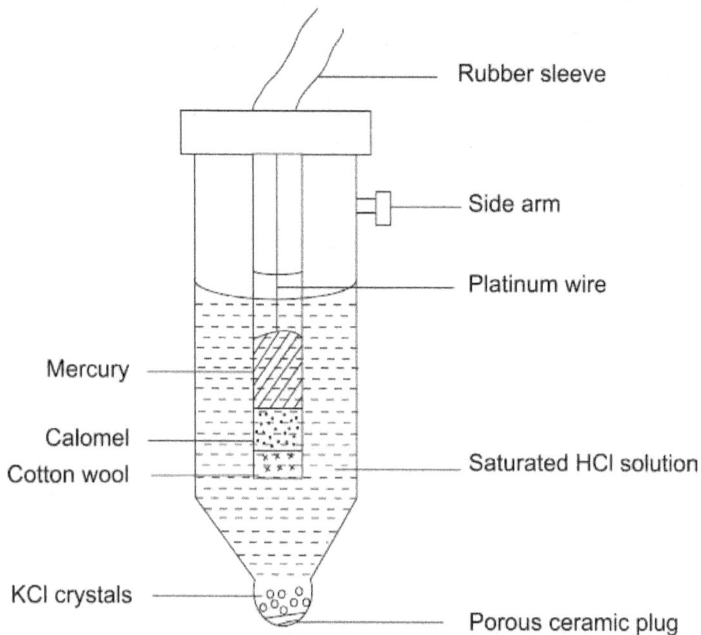

Figure 5.3. Structure of calomel electrode.

The potential developed in the electrode will be always constant. Though other reference electrodes are used in pH meters, the calomel reference electrodes are most widely used because they are cheap and easy to maintain and the results are reproducible and constant.

Glass electrodes   A glass electrode acts as a membrane interface for hydrogen with a thin glass bulb allowing only hydrogen ions. A Typical glass electrode consists of a piece of silver coated with silver chloride, dipped into 0.1N hydrochloric acid contained in a sealed glass tube. The bottom of the tube is a glass bulb, which is very thin and pH-sensitive. (Figure 5.4).

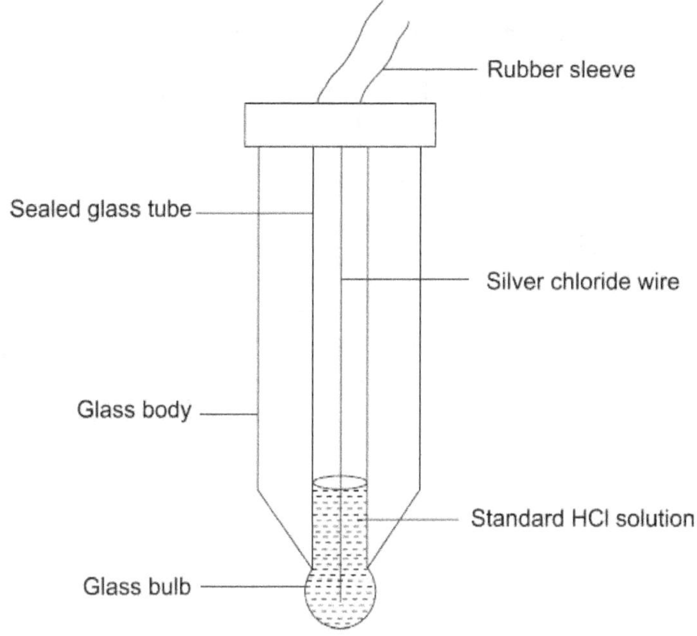

Figure 5.4. Structure of a glass electrode.

General purpose glass electrode   It is used to measure pH of acids, alkalis and organic solutions for the ranges of pH 0–11.

Amber glass electrode   It is made up of an amber glass bulb and is used to measure the pH of solutions in the range between 0 and 4.

E-2 glass electrode   It is a specially designed glass electrode to measure high pH.

Sleeve type reference electrode   It is used to measure pH in highly viscous samples.

Combined electrode   It has both glass and reference electrodes, which are assembled as a single unit.

Flat-ended glass electrode   It has a flat surfaced bottom and is used for pH measurements on flat or moist surfaces.

Micro-spear electrode   It has spear-shaped bottom and is used for pH measurements in small samples.

Oxygen electrode   It is a gas electrode used for measuring the partial pressure of oxygen in a solution. It includes blood gas electrode and Clark electrode.

Blood gas electrode   It consists of a piece of platinum wire embedded in a glass tube with its tip exposed to outside. (Figure 5.5). The electrode is dipped into an electrolyte contained in a chamber into which oxygen is allowed to bubble.

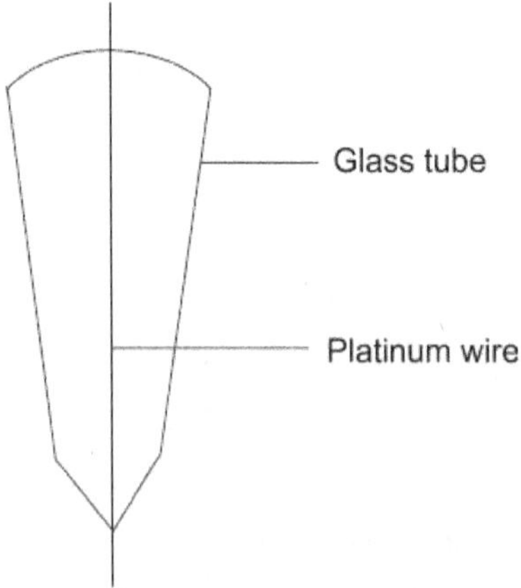

Figure 5.5. A blood gas electrode.

When current is applied, oxygen undergoes reduction at the platinum cathode. The oxidation–reduction potential with reference to the partial pressure of diffused oxygen is measured.

Clark electrode   It is an electrode in which the blood gas electrode (oxygen electrode) and the reference electrode (silver/silver chloride electrode) are combined as a single unit as shown in Figure 5.6. The oxygen electrode with platinum cathode and the reference electrode with silver/silver chloride anode are dipped into buffered potassium chloride solution (electrolyte) contained in a sealed chamber, the bottom of which is covered by oxygen-permeable membranes of polypropylene.

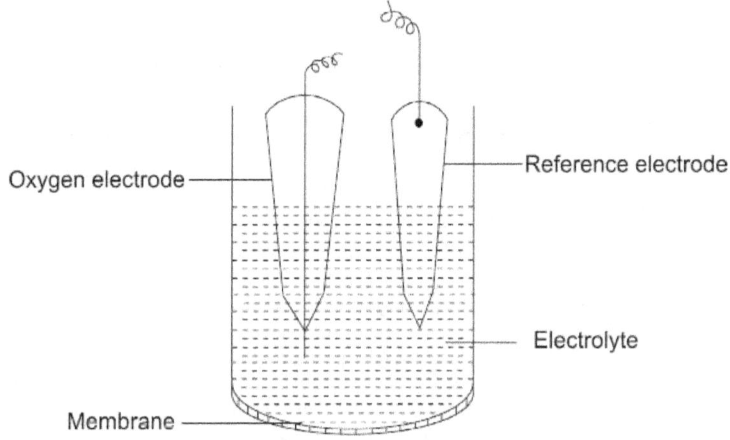

Figure 5.6. Structure of a Clark electrode.

When the electrode is dipped into the sample, oxygen from the sample diffuses into the electrolyte and the electrons are transferred from cathode to anode producing a potential. It is measured and is proportional to the partial pressure of the sample.

Carbon dioxide electrode   It is an electrode in which a glass electrode and calomel reference electrode are combined. The two electrodes make contact with a thin plastic or teflon membrane which is permeable only to carbon dioxide.

Standard bicarbonate solution (0.005 M) is kept between the membrane and the glass electrode as in Figure 5.7. When the electrode is dipped into the sample, carbon dioxide diffuses into the bicarbonate solution through the membrane so that the pH of the bicarbonate solution changes. The pH meter measures this alteration through glass electrode.

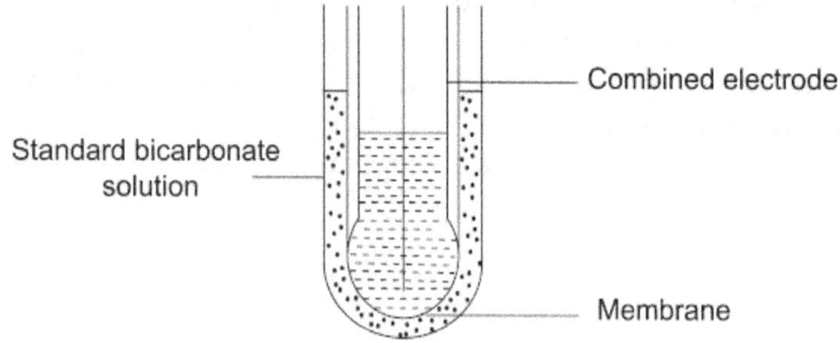

Figure 5.7. Structure of a carbon dioxide electrode.

5.2.3 Factors Affecting Measurement of pH

pH meter is a potential meter consisting of an electrode system and an electrometer which measures even very small differences of electrical potentials in a current circuit. When the glass and reference electrodes are simply dipped into the sample contained in a beaker and the electric current is passed, the pH of the sample is directly recorded from the potentiometer. However, the following factors could interfere the measurement of pH of the samples.

Ionic strength   In a true sense, the pH is actually due to the activity of H^+ ions and is not concerned with H^+ ion concentration. In highly concentrated solutions, the activity of H^+ ions is lower than their concentration. In addition, other ions present in the solution can alter the characteristics of H^+ ions. Therefore, the pH measurement in highly concentrated solutions should be avoided.

Electrode contamination   In glass electrodes, the glass membranes have minute pores through which H^+ ions pass. If these pores are blocked, the permeability of the H^+ ions is reduced especially when the pH of proteins are measured. Therefore, the electrodes must be washed frequently with acid or detergents. Moreover, certain buffers bring alterations in pH value through changing the permeability of glass membranes of the electrodes. Such buffers should not be used.

Sodium error   As the glass electrodes are also permeable to sodium ions, they produce a potential for Na^+ ions also. As a result, the glass electrodes cannot measure the exact pH of samples containing Na^+ ions. Using sodium-impermeable glass electrodes or using potassium hydroxide solution instead of sodium hydroxide in the electrodes can solve this problem.

5.3 IMPORTANCE OF BUFFERS IN BIOLOGICAL SYSTEMS

In living organisms, water forms the main constituent of cell cytoplasm and extracellular fluid. Any aqueous solution is characterized by having positively charged hydrogen ions (H^+) and negatively charged hydroxyl ions (OH^-). The concentration of hydrogen ions in a solution represents its pH. Though the cytosol of the cell has a pH of about 7.2, the pH of cytoplasmic organelles varies. Therefore, the normal functioning of cellular structures requires specific pH maintenance. In general, living organisms have to maintain the pH ranging from 6.5 to 8.0 inside the cells and body fluids. This is because the biological structures are greatly affected by changes in pH. Most of the biomolecules are acidic, basic or amphoteric. If any change occurs in the pH of biomolecules, the biological activity is greatly affected.

A buffer solution is one, which will resist to changes in pH on addition of small quantity of acid or alkali. Generally, a buffer system consists of a weak acid and its salt or a weak base and its salt. In biological systems, the major buffers in extracellular fluids are bicarbonate buffers, phosphate buffers and protein buffers. The main intracellular buffers are phosphate and protein buffers.

5.3.1 Bicarbonate Buffer System

This is the main buffer system in blood plasma and consists of bicarbonate (HCO_3^-) ions and carbonic acid (H_2CO_3). This system metabolizes stronger dietary and metabolic acids (HA) and converts them into weak bases (A) with an increase in carbonic acid concentration. Stronger bases (B) are converted into weak acids (BH^+) with an increase in bicarbonate ions as shown below.

$$HA + HCO_3^- \rightleftharpoons A^- + H_2CO_3$$

$$B + H_2CO_3 \rightleftharpoons BH^+ + HCO_3^-$$

The pH of blood is maintained at 7.4 when the buffer ratio becomes 20. If the bicarbonate buffer neutralizes any acid or base, there may be a change in the buffer ratio and pH value. But the buffer ratio is kept constant by the elimination of carbonic acid as carbon dioxide through respiration or as bicarbonates through urine.

5.3.2 Phosphate Buffer System

The buffering capacity of phosphate buffer is much lower than the bicarbonate buffer system. It consists of dibasic phosphates (HPO_4^{--}) and monobasic phosphates ($H_2PO_4^-$) with pH value of 6.8 in which the ratio of HPO_4^{--} and $H_2PO_4^-$ in plasma is 4. The pH of intracellular fluid is nearer to the pK values of phosphate buffer so that the buffering capacity of phosphate buffer is high inside the cells as well as in renal tubules and collecting ducts of the kidney. When the ratio of monobasic and dibasic phosphate is altered by the formation of more dibasic phosphates, $H_2PO_4^-$ are eliminated so that the ratio becomes unaltered. In biological solutions, NaH_2PO_4 (sodium dihydrogen phosphate) acts as a weak acid and produces H^+ ions in solution.

$$NaH_2PO_4 \rightleftharpoons Na^+ + H_2PO_4^- \rightleftharpoons H^+$$
$$+ HPO_4^{--} \rightleftharpoons H^+ + PO_4^{---}$$

Disodium dihydrogen phosphate ($Na_2H_2PO_4$) acts as phosphoric acid forming a strong base. The addition of an alkali to this buffer system results in the formation of water.

$$NaH_2PO_4 + NaOH \rightarrow Na_2HPO_4 + H_2O$$

The addition of an acid results in the formation of the salt.

$$Na_2H_2PO_4 + HCl \rightarrow NaCl + NaH_2PO_4$$

5.3.3 Protein Buffer System

It is an important buffer system in the plasma and intracellular fluids. Proteins occur as anions serving as conjugate bases (Pr^-) at the blood pH of 7.4 and as conjugate acids (HPr) accepting H^+ ions.

5.3.4 Amino Acid Buffer System

Amino acids possess both basic ($-NH_2$) and acidic ($-COOH$) groups so that the addition or removal of H^+ ions to a zwitter ion amino acid results in the formation of cationic or anionic forms of amino acids as shown below.

$$H_3N^+-CH_2-COOH \rightleftharpoons H_3N^+-CH_2\ COO^- \rightleftharpoons H_2N-CH_2^-\ COO^-$$

$$\uparrow \qquad\qquad\qquad \downarrow$$

$$N^+ \qquad\qquad\qquad H^+$$

(Cation form)   (Zwitter ion form)   (Anion form)
(Glycine)

5.3.5 Buffer System in Human Blood

The buffer system of blood in man includes bicarbonates, carbonic acid, plasma proteins and phosphate mixture of sodium. In tissues, oxidative processes produce carbon dioxide. A major portion of this CO_2, on combination with water, is converted into carbonic acid in red blood cells resulting in increased production of H^+ and HCO^{3-} ions. The bicarbonate ions diffuse into the plasma and combine with Na+ ions to form sodium bicarbonate ($NaHCO_3$). The formation of basic bicarbonates maintains constant pH in the plasma. When there is high tension of oxygen in blood, more oxyhaemoglobin molecules are formed and bicarbonates are used up. As a result, the bicarbonate ions are drawn into the red cells from the plasma. Under high tension of CO_2 in blood, more bicarbonate ions are produced to make up for the ones lost. When bicarbonate ions pass out of the red cells, they are replaced by chloride ions (Cl^-) from plasma (chloride shift). In lungs, opposite reactions occur as in the Figure 5.8.

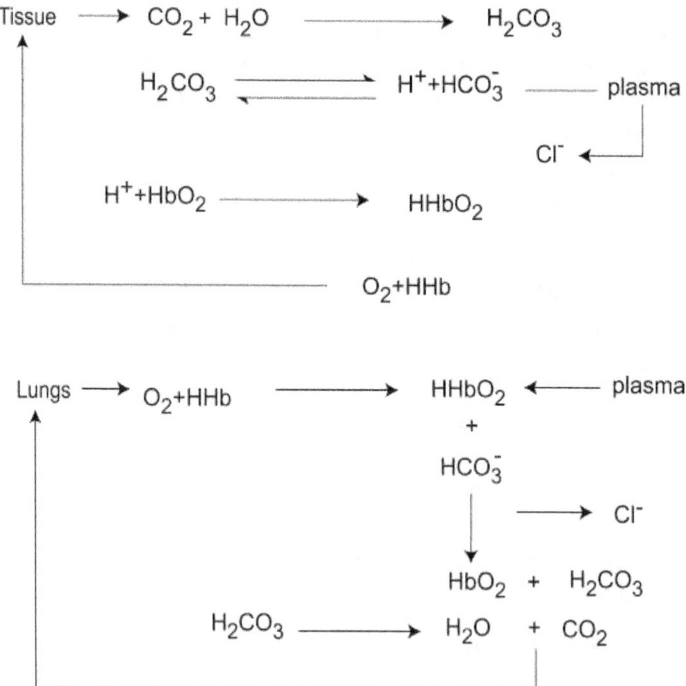

Figure 5.8. Buffer action in human blood.

5.3.6 pH-dependent Ionization of Biomolecules

Ionization of amino acids   Amino acids are amphoteric in nature. A weak basic amino group and an acidic carboxyl group form each amino acid. Therefore, their ionization in an aqueous solution depends on the pH of the solution. In general, the amino acids exist in three ionic states namely cation, zwitter ion and anion. At lower pH, an amino group amino acid undergoes ionization, making the amino acid the cation. When pH is increased, the carboxyl group of an amino acid begins dissociation, making the amino acid negatively charged. During this process, the charge is cancelled and the amino acid has no change and is called Zwitter ions. Further increase of pH causes the amino acid to dissociate, leaving a negative charge on the amino acid. Thus the amino acid becomes an anion. The whole process of ionization of an amino acid is given in figure 5.9.

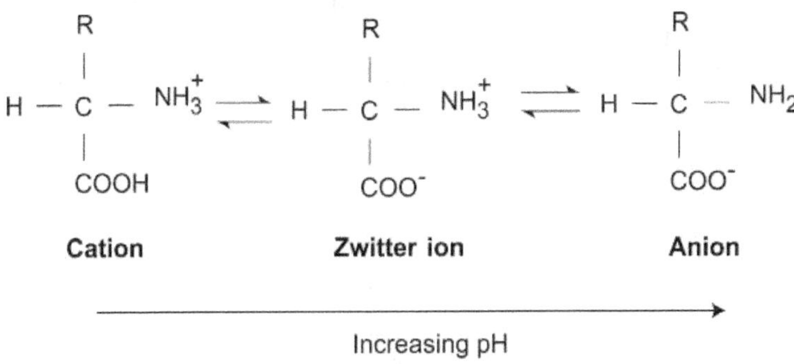

Figure 5.9. Ionic forms of an amino acid depending on pH.

Consider that a solution of pH 5.0 contains a mixture of three amino acids. Under an electric field, the amino acid with pH of 5.0 (A) will be static without any charge and so it is in zwitter ion state. The amino acid (B) with pH of 3.50 will become negatively charged and move towards the anode. The amino acid (C) with pH of 8.20 will become positively charged and move towards the cathode as in Figure 5.10.

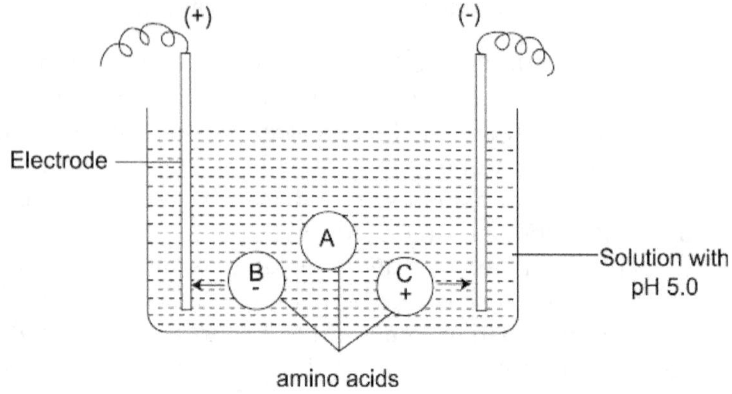

Figure 5.10. pH-dependent ionization of amino acids.

Thus ionization of amino acids depends on the existing pH and so a mixture of amino acids can be separated by electrophoresis as they posses different potential differences at different pH. Moreover, the amino acids also exhibit pH-dependent differential solubility. For example, at isoionic pH, the solubility of amino acids is lesser but greater at lower and higher pH.

Ionization of proteins   As proteins are formed by amino acids through peptide linkage, the ionization of proteins also depends on pH. In a protein molecule, free amino groups at N-terminal end and free C-terminal end are available for ionization as in the Figure 5.11. Moreover, the side chains in a protein molecule also undergo ionization, which is also pH dependent. Therefore, separation of protein molecules can also be accomplished by electrophoresis.

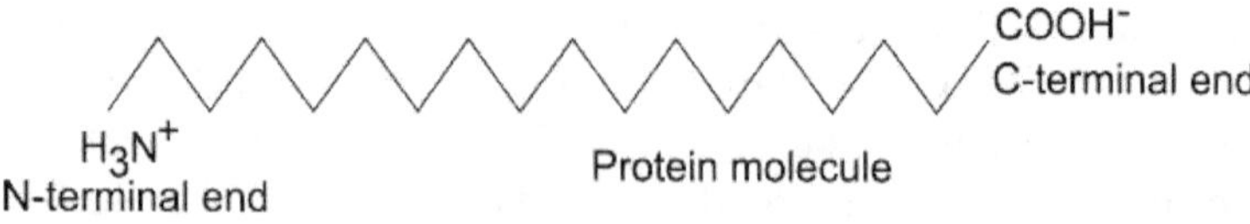

Figure 5.11. N– and C– terminal ends of amino acids in a protein molecule.

5.4 CENTRIFUGATION

When a solution contained in a vessel is rotated at high speed, its particles will sediment due to centrifugal and gravitational forces. The movement of homogenous solutes as a sharp boundary is called **sedimentation.** It is the transport of matter in a mixture by external force particularly gravity or centrifugal force and is expressed as Svedberg unit 'S'. One 'S' is equal to 1×10^{-18} cm/sec/dyne/g. The basic principle of sedimentation is the separation of homogenous particles from a mixture based on their size or density by centrifugation with required time and particular centrifugal field. The rate of movement of particles depends on the centrifugal force, size, shape and density of the particles and the density and viscosity of the medium in which the particles are suspended. The **centrifugal force** depends on the angular velocity of the centrifuge and the distance of the particles from its axis. The asymmetric particles sediment at a slower rate than the spherical particles as they face the problem of friction. The frictional ratio for spherical particles is closer to 1 whereas it is larger for asymmetric particles. Thus two particles which have the same weight but different shapes, will sediment at different rates.

The **sedimentation coefficient** is the velocity of the particles to settle down per unit centrifugal field. The **rate of sedimentation** or **sedimentation velocity** is increased with the concentration of a solution, which in turn depends on the size, and symmetry of particles. For macromolecules, the rate of sedimentation is not proportional to the centrifugal force, so they aggregate to form macromolecular clusters at high speed and settle down. But the sedimentation of ionized macromolecules is much quicker than unionized molecules.

5.4.1 Basic Principle of Centrifugation

When an object is moved in a circle at a steady angular velocity, it will experience an inwardly directed force F. The magnitude of F is determined by the angular velocity (ω) and the radius of rotation (r) and is represented as

$$F = \omega^2 r$$

In order to express F value in terms of gravitational force, the value is divided by 980 and denoted as **Relative Centrifugal Force** (RCF). The RCF depends on the number of revolutions per minute (rpm) (ω), radius of rotation (r) and gravitational force (g).

Then,

$$RCF = \frac{\omega^2 r}{980}$$

The radius (ω) and rpm can be related by the following formula:

$$\omega = \frac{\pi \, (rpm)}{30}$$

Therefore,

$$RCF = \frac{\dfrac{\pi^2 \, (rpm)^2}{30^2} \, r}{980}$$

The revolutions per minute and gravitational force can be related by the equation,

$$G = \omega^2 r$$

where,

$$\omega = \frac{\pi \, (rpm)}{30}$$

Therefore

$$G = \frac{\pi^2 (rpm)^2}{30^2} r$$

where,

r = radial distance from the axis of rotation.

The particles will get increased RCF values during every successive period of centrifugation due to a change in radius every time. So the radius of a centrifuge tube is to be measured from the rotor axis to the middle of the liquid column (average radius) as shown in the Figure 5.12.

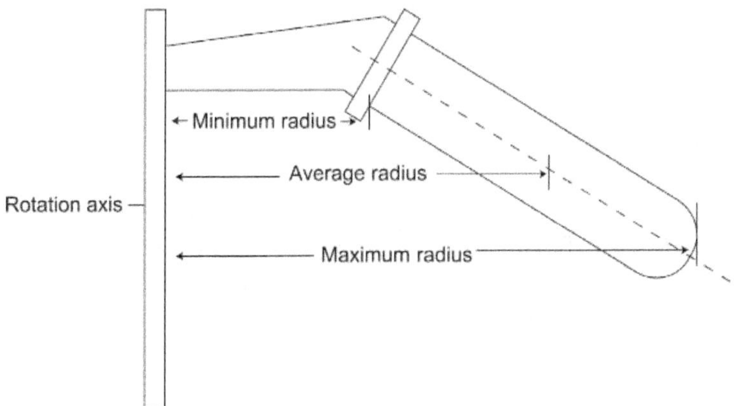

Figure 5.12. Radii at different regions of a centrifuge tube.

CENTRIFUGES

A centrifuge is a device to separate homogeneous particles in a mixture according to their size, shape and density. Typically, a centrifuge consists of a metal rotor with a provision to accommodate the sample vessels, a motor to rotate the rotor with desired speed and accessories. Based on the speed of centrifugation, the centrifuges are categorized into the following types.

5.5 DESKTOP CENTRIFUGE OR CLINICAL CENTRIFUGE

It is small and simple apparatus with a speed up to 3000 rpm but without temperature-regulating accessories. It is generally used in clinical works for rapid sedimentation of substances.

5.6 HIGH-SPEED CENTRIFUGE

It has a speed up to 25000 rpm and is provided with a cooling mechanism and temperature regulator. It is widely used to sediment microorganisms, cells, large cellular organisms, etc.

5.7 ULTRACENTRIFUGE

It consists of a titanium or aluminium rotor to which the centrifuge cells are held. The rotor is rotated by an electric motor to a speed up to 70000 rpm. The entire motor assembly is kept inside a heavy evacuated and refrigerated chamber in order to avoid excess heat. To balance the weight on the rotor, a flexible shaft is used. Optical devices are provided to monitor

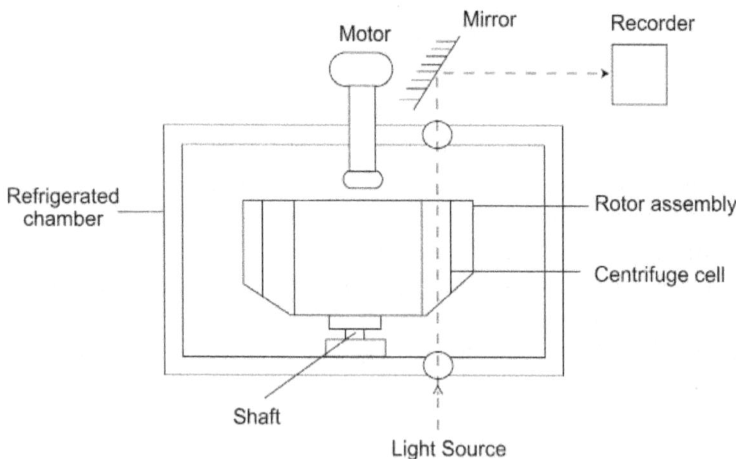

Figure 5.13. Schematic structure of an ultracentrifuge.

the movement of the boundary during centrifugation and to record the refractive index of the particles. A schematic structure of an ultracentrifuge is given in Figure 5.13.

Ultracentrifuge includes two types namely preparative ultracentrifuge and analytical centrifuge.

Preparative Ultracentrifuge

It has a cooling system to maintain the temperature between 0°C to 4°C with a speed up to 75000 rpm. The rotor system is completely enclosed in an evacuated armoured chamber in order to minimize the heat and to avoid explosion. The shaft is made flexible to avoid spindle damage and to withstand the centrifugation force. It is also provided with speed regulator.

This centrifuge is useful to study structural and functional aspects of subcellular organelles and to characterize macromolecules such as proteins, DNA, RNA, etc.

In general, preparative centrifugation is mainly concerned with isolation of biological materials for biochemical analysis. Based on the medium of suspension used, preparative centrifugation is of two types namely **differential centrifugation** and **density gradient centrifugation**.

5.7.1 Differential Centrifugation

If the separation is done in a homogeneous suspending medium, it is called differential centrifugation. This method is mainly useful for cell fractionation (isolation of intracellular organelles). In recent times, the fractionation of cells has been accomplished by cellular disintegration and differential centrifugation in a suitable medium.

Cell fractionation   The tissue is homogenized in an isotonic medium (0.25 M sucrose solution or 0.9 % sodium chloride solution) to disrupt plasma membrane of cells with the help of a homogenizer. The homogenate is filtered to remove unbroken cells and debris and the cellular organelles are fractioned by centrifuging the homogenate at increasingly higher speeds. The process of cell fractionation by differential centrifugation method is shown diagrammatically in Figure 5.14. When the homogenate is centrifuged at low speed, the nuclei are sedimented. Mitochondria, lysosomes and peroxisomes are sedimented when the supernatant is centrifuged at higher speed. Subsequent centrifugation of the supernatant at still higher speed results in the deposition of plasma membrane, microsomes and large polyribosomes. At the highest speed, centrifugation of supernatant causes sedimentation of small ribosomes. By centrifugation, of the supernatant at the highest speed for a long period, the cytosol, the soluble aqueous fraction of cytoplasm, is obtained.

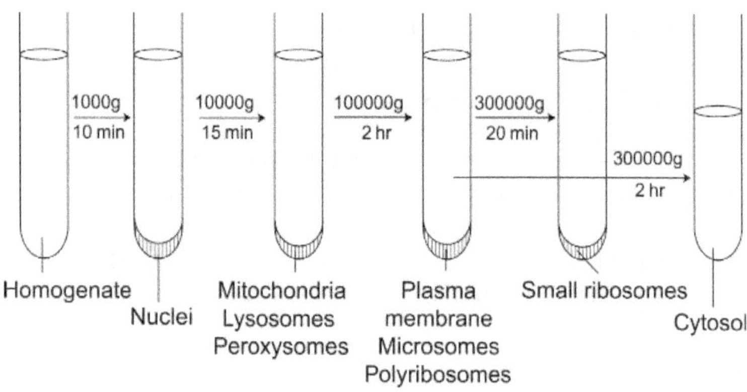

Figure 5.14. Fractionation of a cell homogenate.

In the above method, all cellular particles are homogeneously distributed in the homogenate, prior to centrifugation. The particles are sedimented according to their sedimentation rate by centrifugation.

5.7.2 Density Gradient Centrifugation

If the separation is done in a medium having different density gradients, then it is called density gradient centrifugation.

Selection of gradient medium   The selection of gradient medium is an important prerequisite to separate particles by gradient centrifugation. This is because the gradient medium should not affect the sample and should be easily sterilizable, recoverable and non-corrosive. The most commonly used gradient media include sucrose, glycerol, sorbitol, etc.

Two types of gradients namely discontinuous and continuous density gradients can be made in the medium. In discontinuous density gradient medium, the density increases from one layer to another. This medium can be prepared by pouring solutions of decreasing densities one over each other with the solution of highest density at the bottom. This medium is useful in the separation of whole cells, subcellular organelles or lipoproteins. In continuous density gradient medium, the density decreases from the bottom of the solution to the meniscus. This medium can be prepared by keeping the solution undisturbed for a long time so that layers of linear continuous gradient are formed. This medium is useful in the separation of ribosomes, viruses, proteins and enzymes.

Density gradient centrifugation includes two types namely **rate-zonal centrifugation** and **isopycnic centrifugation**.

Rate-zonal centrifugation   By this method, centrifugation is carried out at a very low speed for a short time so that the particles settle down. However, centrifugation should be stopped before the particles of any zone settle at the bottom. This type of centrifugation is useful in the separation of nucleic acids and ribosomal subunits.

Isopycnic or sedimentation equilibrium centrifugation Here, the particles of a solution move according to their buoyant densities and become static at a place where the density is greater than their own. This requires a very long time and high speed. As an example, the pellet obtained by centrifugation of the tissue homogenate at 10000 g is suspended in increasing densities of sucrose solution and centrifuged for several hours at 40000 rpm. Now the individual organelles move to the region of their own equilibrium density and remain at specific regions as shown in Figure 5.15. This method is useful in the separation of proteins, intracellular organelles and nucleic acid fractions.

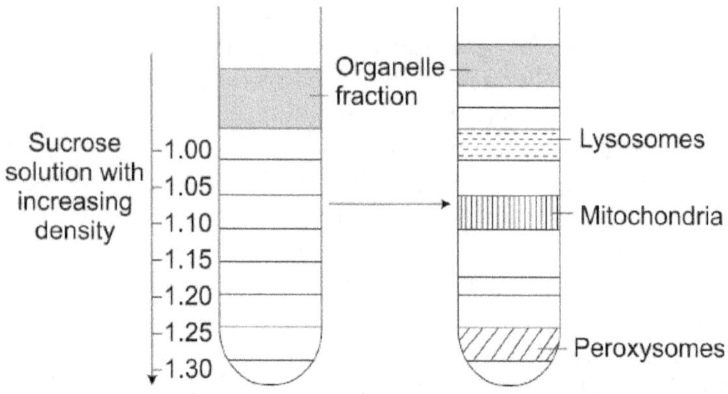

Figure 5.15. Separation of organelles by sedimentation equilibrium centrifugation.

Analytical Ultracentrifuge

This instrument is similar to that of the preparative ultracentrifuge except for the following differences. The rotor in this centrifuge is elliptical in shape with two holes to hold centrifuge tubes of which one is analytical and the other is for counter balancing. The holes are calibrated to measure the distances in the analytical tube. The upper and lower ends of the tubes are transparent and are formed by quartz windows. These windows are meant for monitoring the progress of centrifugation. The optical system is capable of recording the sedimentation of particles continuously and the entire process of sedimentation can be photographed. This centrifuge is useful to determine the sedimentation coefficient and the molecular weight of macromolecules.

5.7.3 Determination of Molecular Weight by Ultracentrifugation

By using ultracentrifugation technique, the molecular weight of proteins, nucleic acids and enzymes can be determined by two methods namely **sedimentation velocity method** and **sedimentation equilibrium method**

Sedimentation velocity method When a solution containing uniformly-distributed solutes is centrifuged at high speed (55000 rpm), the particles migrate outwards from the centre of rotation, forming a well-defined boundary between the solvent portions with and without particles. The velocity of the boundary is measured by using optic systems (Schlieren or Rayleigh interference optics) provided in the centrifuge. From this value, the sedimentation constant (Z) can be obtained. The molecular weight of the particles can be determined by the equation,

$$M = \frac{RTZ}{D\,(1-V\rho)}$$

where,

   M = molecular weight

   R = gas constant

   T = temperature

   D = diffusion coefficient

   V = volume of the substance and

   ρ = density of the medium

Sedimentation equilibrium method   In this method, the centrifuge is run at low speed until the particles stop movement. At this point, the sedimentation velocity at which sedimentation is obtained balances the diffusion of particles. The concentration gradient formed in the centrifuge tube is measured and the molecular weight of the particles is determined by using the equation,

$$M = \frac{2RT\, I_n\, C_2/C_1}{\omega^2\,(1-V\rho)\,(X_2^2 - X_1^2)}$$

where,

C_1 and C_2 = concentration of substances at distances
X_1 and X_2

ω = angular velocity of the centrifuge.

5.8 CHROMATOGRAPHY

It is a perfect analytical technique for the identification, isolation and separation of compounds based on differences in affinity for a stationary phase and a mobile phase. When a solution is applied to an insoluble medium having differential affinity for individual molecules of solution, the molecules will migrate through the medium at different rates. In chromatography, the stationary phase may be a solid, liquid, gel or a solid/liquid mixture and the mobile phase may be a liquid or a gas.

5.8.1 Principles of Chromatography

The basic principle of chromatography is adsorption and partition. When a mixture of compound is allowed to pass through stationary and mobile phases, the compounds become separated according to their Rf (Resolution front) values.

$$Rf = \frac{\text{Distance travelled by the sample}}{\text{Distance travelled by the solvent front}}$$

In all the chromatographic systems, the differences in affinity involve either adsorption or partition. In adsorption, a compound binds to the surface of the solid phase whereas in partition, the relative solubility of a compound in two phases results in the partition of the compound. In chromatographic separation of compounds, most of the interactions is physical in nature and chemical reactions also play a role. The physical interactions include nonpolar interactions as well as interactions involving hydrogen bonds. The chemical interactions include ion exchange (ion–ion interaction) and the affinity between a macromolecule and a ligand.

The interactions during chromatographic separation of substances include the following:

Partition   When a solute is dissolved in two immiscible liquids of equal volume, the ratio of its concentration in two phases at a given temperature and equilibrium is called **partition coefficient**. The partition chromatography and gas-liquid chromatography are mainly based on this interaction. Here the relative solubility of the compound influences its movement as the solvent travels along the stationary phase. Therefore, the movement of more soluble substances in the mobile phase is greater than the migration of substances, which are more soluble in the stationary phase.

Adsorption   Separation of substances can be achieved by differential adsorption of substances by the mobile and stationary phases. Here the interactions take place between the hydrogen bonds and electrostatic forces of the substance and the adsorbent. As the interacting solutes with the adsorbent are retarded more when compared to less interacting solutes, the compounds of the sample are separated.

Ion exchange   It is the process of reversible exchange of ions in solution with the ions, which are bound to the insoluble medium. The ion exchange substances such as cellulose contain either cationic or anionic groups, which exchange ions reversibly with other ions in the surrounding medium. Thus cationic exchangers are used for the separation of anions, and anion exchangers for cations. In this way ionic exchange materials act as static phase and electrolyte solution as mobile phase.

Gel-filtration   When a gel containing pores is used as a supporting medium or aqueous stationary phase, the molecules with smaller size than the pore size move slowly whereas the bigger molecules move rapidly. In this way, the molecular size of the particle forms the basis for the separation by gel-filtration.

Affinity of substances   When the sample is passed through a substrate in combination with a gel matrix, the molecules which are specific for the substrate get bound to the gel and become immobile whereas the other molecules move freely.

TYPES OF CHROMATOGRAPHY

The chromatographic technique is of different types based on various principles described above. A brief account of important chromatographic methods is given below.

ADSORPTION CHROMATOGRAPHY

An **adsorbent** is a solid capable of holding molecules at the surface without requiring electrostatic forces. Cellulose, starch, silica gel, alumina, etc. are commonly used as adsorbents. In this technique, any organic solvent such as carbon tetrachloride, benzene, chloroform, ether, acetone, ethanol, water, etc. are used as mobile phase. Adsorption chromatography includes **paper chromatography, thin layer chromatography and column chromatography.**

5.9 PAPER CHROMATOGRAPHY

It is the simplest and widely used method in many disciplines of science. In paper chromatography, the substance is analysed by allowing the flow of solvents on a specially designed filter paper. The solvent rises up on the paper by capillary action and the separation is effected by differential migration of the substances of a mixture. In **paper partition chromatography,** paper is used as immobile phase and the solvent as mobile phase. In **paper adsorption chromatography**, the paper is impregnated with an adsorbent like alumina and the solvent is allowed to flow over the paper.

5.9.1 Choice of Paper

In paper chromatographic technique, Whatman paper I which is available in different grades is extensively used. The choice of the paper depends on its thickness and the flow rate, purity and strength of the solvent. To get desired properties, the paper can be impregnated with aqueous buffer solution or its cellulose structure can be modified. The size and shape of the paper depends on the nature of separation involved. Though square or rectangular papers are generally used, rectangular papers are preferred.

5.9.2 Application of the Sample on the Paper

The sample is prepared by using proper solvent and is applied on the paper as spots or bands. A pencil line is drawn on the paper of a suitable size at a distance from the bottom, leaving some distances at the sides. A few points at equal distance are marked on the line as shown in the Figure 5.16.

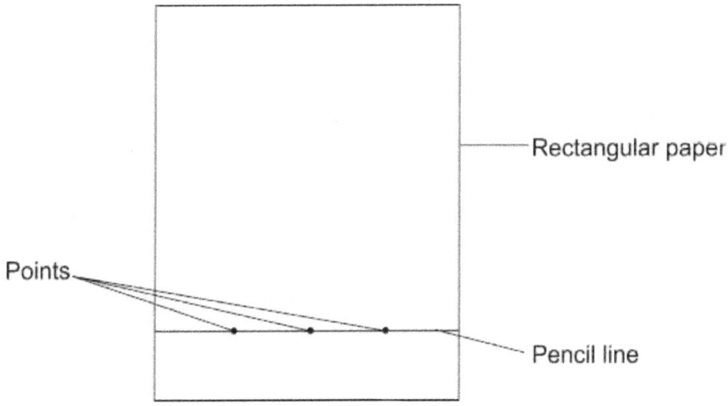

Figure 5.16. Preparation of paper for chromatography.

The sample is applied in drops on the spots with the help of a capillary tube or a microsyringe or a platinum loop. The spotting is done slowly and carefully in such a way that the spots occupy only small areas.

5.9.3 Selection of Solvents

The number of solvents to be used depends on the nature of the substances to be separated. The use of minimum

Table 5.4. Solvent systems to detect substances.

Substance	Components of solvents	Ratio
Amino acids	n-butanol–ethanol–water	4:1:1
	n-butanol–acetic acid–water	4:1:5
	phenol–water	80/20 v/v
	alcohol–water–ammonia solution	20:2.5:2.5
Sugars	n–butanol–acetic acid–water	4:1:5
Lipids	chloroform–benzene–ethanol	48.5:48.5:3.0
	chloroform–methanol–water	70/22/3 v/v/v
Organic acids	n–butanol–formic acid–water	10:2:15
Group I metallic ions	distilled water or ethanol	—
Group II metallic ions	ethanol–water–1N HCl	22.5:1.25:1.25
	Saturated n–butanol with 1N HCl	—
Group IV metallic ions	acetone–water–conc. HCl	21.75:1.25:2.0

number of solvents is preferable. When more than one component of the solvents is mixed together, two distinct layers would result, so the solvents should be shaken well before the use. Moreover, the solvents should be used in specific ratios to separate different substances as shown in Table 5.4.

5.9.4 Development of Chromatogram

A cylindrical or rectangular glass jar can be used as the chromatographic chamber (a chamber made of materials insoluble in the solvent is chosen). Required quantity of the solvent is taken in the chamber. The spotted paper is made into a cylinder by using threads with a distance left between them. Now the paper cylinder is placed in the solvent contained in the chamber. The spotted line on the paper should remain a little distance above the level of the solvent and the paper should not touch the sides of the chamber. The chamber is closed air-tightly. Now the solvent moves over the spot by capillary action. The chromatogram is run until the solvent reaches the top edge of the paper. At this point, the paper is removed and the solvent front is marked. The chromatographic setup is shown in the Figure 5.17.

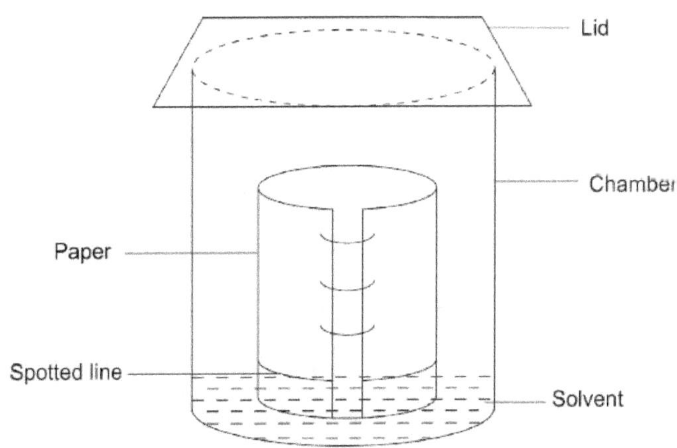

Figure 5.17. Diagram showing chromatographic setup.

5.9.5 Location of Spots

The developed paper is kept in an oven or on a hot plate for few minutes for drying. The paper can be dried with the help of a fan or a hair dryer. It can also be air-dried over night. If the spots are coloured, there would be no difficulty in locating them. To identify colorless spots, there are several methods, which may involve physical or chemical means. In physical methods, ultraviolet light and radioactive elements are used. In chemical methods, the colourless spots are made coloured by reagents called **locating reagents** which may be gases, liquids or solids. The locating reagents for various substances are given in Table 5.5. The locating reagent is applied on the developed chromatographic paper by spraying the reagent on the paper or dipping the paper in the reagent. Then the paper is dried at room temperature overnight or in an oven at 110°C for 10 minutes. Now the spots are marked on the paper. Rf values are calculated and compared with standard Rf values.

Table 5.5 Locating reagents to detect substances.

Substances	Locating reagents
Amino acids	0.1% ninhydrin in acetone or n-butanol
Sugars	0.5% benzidine mixture
Lipids	0.1 sudan black
Organic acids	0.05% bromocresol green
Group 1 metallic ions	0.25M potassium chromate solution
Group II and IV metallic ions	10 ml rubeanic acid + 10 ml ethanol

The method described above is **ascending paper chromatography** in which the solvent flows upwards on the paper by capillary action. In **descending paper chromatography,** the solvent is allowed to flow down by capillary action and gravitational pull on the paper by keeping the solvent at the top of the chromatographic chamber. The above method forms **unidimensional** or **one-dimensional chromatography**. Here the Rf values of substances depend on the size, nature of the substances, solvent systems, diffusion coefficient and so on. Therefore, substances move on the paper at different rates. The substances, which have very close or nearly same Rf values, cannot be separated by the uni-dimensional chromatography. This problem is overcome by **two-dimensional paper chromatography.**

Two-dimensional Paper Chromatography

Two pencil lines are drawn on a square paper of suitable size at distances from the bottom and left margins. The sample is applied on the spot marked 'O' near the pencil lines meet. The paper is developed as in uni-dimensional chromatography in the direction indicated by arrow mark as shown in Figure 5.18, using solvent system I. Then the paper is removed, turned through 90° and developed again using solvent system II. The Rf values are calculated and are compared with standard values after identifying individual spots by using locating reagents.

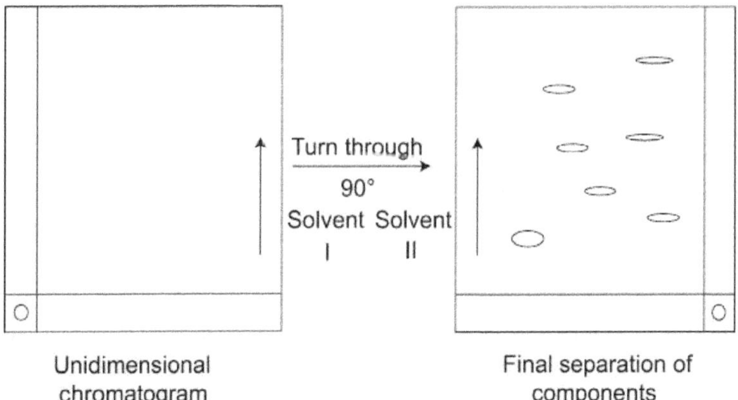

Figure 5.18. Resolution of components in two-dimensional chromatography.

5.10 THIN LAYER CHROMATOGRAPHY (TLC)

The principle and technique of TLC is very similar to paper chromatography. Here the sample is spotted at one end of a glass plate, which is coated, with the slurry of the adsorbents (Sodium acetate or silica gel). Using solvent systems develops the spotted plate and the spots are deducted as in paper chromatography. The adsorbents and solvent systems used to separate substances in TLC technique are shown in Table 5.6.

Table 5.6. Adsorbents and solvent systems in TLC.

Substances	Adsorbents	Solvents	Ratio (v/v)
Amino acids	Silica gel G	Ethanol–water	70/30
		n-Butanol–Acetic acid-water	60/20/20
Sugars	Sodium acetate	Ethyl acetate–n-propanol	65/35
		n-butanol–acetone–phosphate buffer (pH 5.0)	40/50/10
Lipids	Silica gel G	Petroleum ether–diethyl ether–acetone	89/10/1
		Chloroform–Methanol–Water	65/25/10
		Carbon tetrachloride–Chloroform	95/5

The separation of substances by TLC is faster than by the paper chromatographic method. TLC is commonly used to separate components in complex mixtures and the results obtained are more reproducible.

5.11 COLUMN CHROMATOGRAPHY

By using columns, adsorption, partition, ionizing and affinity chromatographic methods can be carried out. Glass or polyacryclic plastics with different dimensions may form the columns used. The selection of the column depends on the quantity of the sample to be separated.

The column is filled with the mobile phase to one-third height by closing the bottom with glass wool or sintered glass. The stationary phase (gel or resin) is made into slurry and is poured into ¾ of the column. By removing the glass wool, the outlet of the column is opened and washing it with the mobile phase stabilizes the column material. Now the sample is added

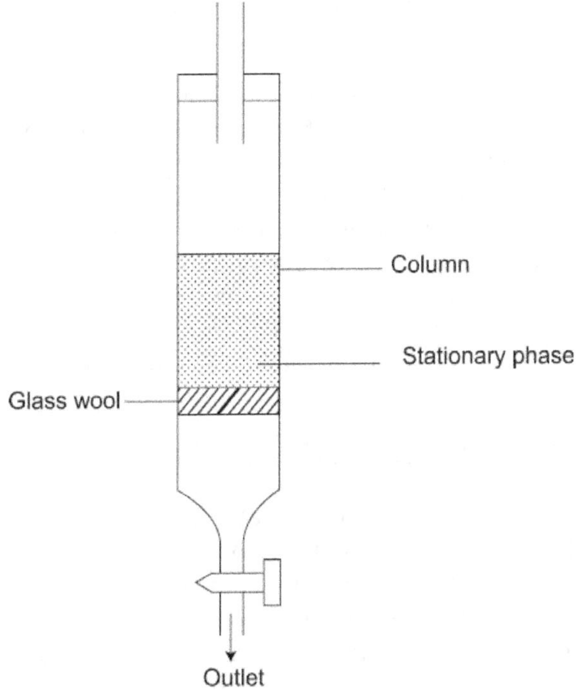

Figure 5.19. Set-up for column chromatography.

onto the surface of the column with the help of a pipette. Above the sample, the solvent is added to a height of 5–10cm. Now the movement of mobile phase through the packed column causes the separation of the components of the sample (column development) (Figure 5.19).

When a single solvent is used as mobile phase, then the separation is called **isocratic separation**, which gives no satisfactory resolution. When two solvents differing in composition are mixed and used as mobile phase, it is **gradient separation** which gives better resolution.

The fractions of the sample which emerge out from the column are collected in different containers and are analysed by using a number of equipments such as fluorescence detectors, polarimeters, voltmeters, refractometers, conductivity meters and so on.

When alumina is used as packing column, the column chromatography works on the principle of adsorption and is called **adsorption column chromatography**. If a column of a gel or modified cellulose is used, partition of substance occurs between the gels and the mobile phase, so the method is called **partition column chromatography.** On the other hand, if the column is packed with an ion-exchange resin, the process of ionic exchange separates the substances and so the method is called **ion-exchange chromatography.**

5.12 GEL PERMEATION CHROMATOGRAPHY

When a gel having pores on its matrix is packed in the column and a heterogeneous mixture is passed through it, smaller ions enter the gel and come out of the column phase. But the bigger molecules cannot enter the gel, so they move only with the mobile phase through the inner phases of the gel. In this process, a type of sieving of molecules occurs and so this method is called **molecular sieve chromatography** or **molecular exclusion chromatography.** As this method involves the use of gel, it is otherwise called as **gel filtration chromatography**. In gel permeation chromatography, the gel which is used as the medium should be highly rigid and chemically inert and should possess small amount of ionic groups with uniform pore sizes. Sephadex, agarose, polyacrylamide, porous glass and silica granules, polystyrene, etc. possess all the above characteristics and so have been widely used. Using the technique of column chromatography or thin layer chromatography gel permeation chromatography can be carried out.

Gel permeation chromatography is widely used in separating and purifying biological molecules. In addition, it is also useful in the separation of low molecular weight compounds such as amino acids, small peptides and oligonucleotides and to remove salts and small molecules from macromolecules.

5.12.1 Determination of Molecular Weight of Macromolecules by Gel Permeation Chromatography

By running the chromatogram, a fraction of macromolecules is collected and its volume is known. From this value, its distribution coefficient is calculated. In a given gel, the distribution of solute particles between inner and outer solvent is referred to as its **distribution coefficient** (K_d). The volume of individual fraction (V_e) depends on the volume of outer solvent (V_o) and volume of inner solvent (V_i)

i.e.
$$V_e = V_o + K_d V_i$$

The volume of the inner solvent can be calculated by the formula

$$V_i = a W_r$$

where,

a = Dry weight of the gel and

W_r = water regain value

Therefore,

$$V_e = V_o + (K_d \times aW_r)$$

Then,

$$K_d = \frac{V_e - V_o}{aW_r}$$

A standard graph with distribution coefficient of standards plotted against their molecular weight is referred, to know the molecular weight of the fractioned compounds.

5.13 ION-EXCHANGE CHROMATOGRAPHY

Separation of substances by ion-exchange chromatography is carried out in columns packed with ion-exchangers (anion- and cation-exchangers), which act as inert and insoluble supporting media. Polystyrene and cellulose are mainly used to prepare ion-exchange resins. In addition, sephadex and sepharose are also useful to separate high molecular weight compounds.

After the removal of impurities, an anion-exchanger is treated with 0.5N HCl followed by 0.1N NaOH in order to expose the charged functional groups in the medium. The reverse of the above treatment is carried out for cationic exchangers. The matrix is then treated with EDTA to remove metal ions and is repeatedly washed in water to remove very small particles. Finally, the exchanger is washed with different reagents to get desired counter ions, the excess of which are washed with large volume of water.

The nature of buffer solution depends on the type of substances to be separated and nature of ionic exchange. For example, an anion exchange requires cationic buffers or vice versa. A suitable pH should be maintained in the buffer. Ammonium acetate (pH 4.6), ammonium formate (pH 3.0–5.0) and ammonium carbonate (pH 8.0–10.0) are some volatile buffers used in ion-exchange chromatography. The quantity of the sample required for analysis depends on the size of the column and the capacity of the exchangers. Gradient separation is more preferred in ion-exchange chromatography than the isocratic separation.

Uses   It is useful in the analysis of amino acids (ion-exchange principle is the basis for the construction of amino acid autoanalyser) to purify water and to detect base composition of nucleic acids. It is also used to detect the concentration of trace metals in biological samples.

5.14 AFFINITY CHROMATOGRAPHY

It is a type of column chromatography in which the column is a water-insoluble carrier or supporting matrix. It should be stable, porous and inert and should possess good flow properties and suitable chemical groups for ligand attachment. The most commonly used matrix includes agarose, polyacrylamide and porous glass beads. Of these, agarose is a polysaccharide support and is most commonly used. The water-insoluble substances are selectively made to combine with the column material by a different substance called ligand. The ligand unites with the column matrix as well as with the macromolecules by covalent linkage.

The agarose support is treated with cyanogen bromide at a pH of 11.0 (which is kept constant by adding 2M NaOH solution) at a constant temperature ($20°C$). The reaction mixture is washed in Tris, sodium bicarbonate or borate buffer. Then the ligand is mixed with activated agarose in the presence of Tris buffer and stirred for a few hours at room temperature and then the matrix is washed in 0.1 M glycine buffer at pH 9.0 to remove extra-activated groups and unbound ligand. The sample is applied at the top of the column and the buffer flow is started. The separated macromolecules flow out of the column and are collected for the analysis.

Uses   It is useful for purification of macromolecules like enzymes, immunoglobulins, polysaccharides nucleic acids, whole cells, etc.

5.15 GAS-LIQUID CHROMATOGRAPHY (GLC)

This technique is another form of column chromatography in which the sample is carried as vapour in an inert gas such as argon, nitrogen or helium over liquids like silicone oils or lubricating greases held by solids like diatomaceous earth or ground firebrick. That is, a non-volatile liquid is the stationary

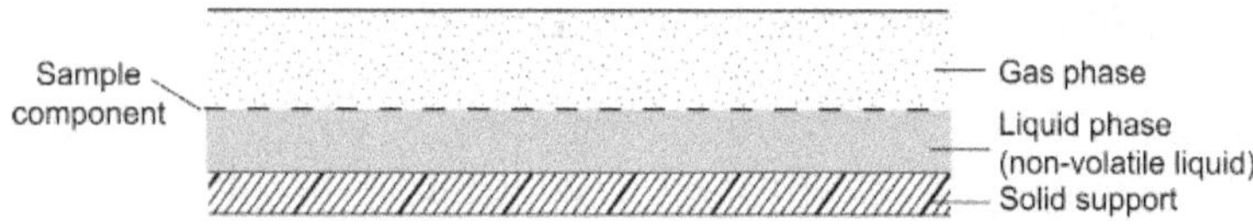

Figure 5.20. Distribution of sample in the column of GLC.

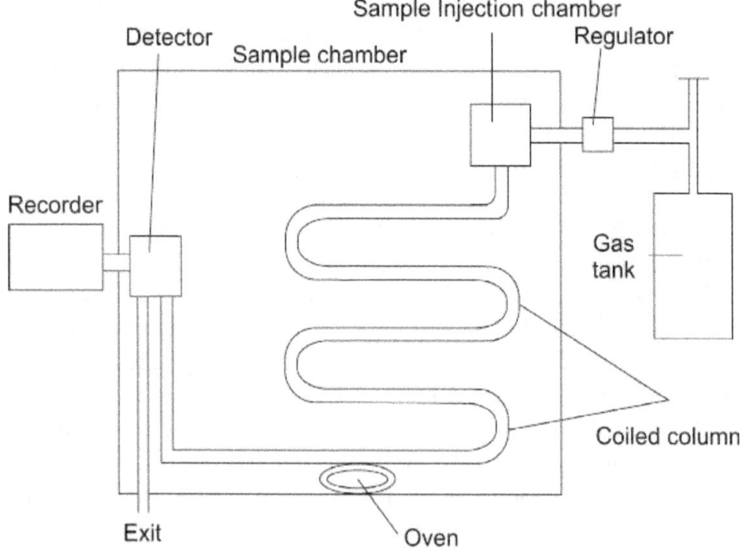

Figure 5.21. Components of GLC (Schematic).

phase, which is dispersed on the surface of solid support (holder of the liquid phase), as shown in the Figure 5.20. The carrier gas stored in a tank is passed into the sample injection chamber. From the chamber, the sample is carried by the gas into the column.

In the column, the components of the sample get distributed between the liquid and gas phases (Figure 5.21) and become separated from each other. The separated components of the sample are now passed onto the detector and then to the recorder. In this method, the separation process requires high temperature and so the instrument is provided with an oven and thermostat.

Uses   GLC is useful in the separation of the components of smoke, atmospheric pollutants, plant extracts, organic acids, etc.

5.16 HIGH PERFORMANCE LIQUID CHROMATOGRAPHY (HPLC)

HPLC is also a column chromatography and is designed to increase the flow rate of the solutes in order to reduce time consumption. The components of a HPLC system are depicted schematically in the Figure 5.22.

The solvent delivery system has an inert reservoir with a provision for degassing the solvent (mobile phase). The selection of mobile phase depends on the type of separation (isocratic or gradient) and the solvents must be extra pure. The pumping system generates high pressure and directs a constant flow of the solvent to the column. The solvents used as mobile phase include ethanol, methanol, propanol and acetonitrate. Pure water or buffers can also be used. The sample is dissolved in the mobile phase and filtered. Using a microsyringe or a metal loop which would withstand high pressure, the sample is injected. The sample can be introduced either directly above the column or to the column, which is made up of stainless steel, glass,

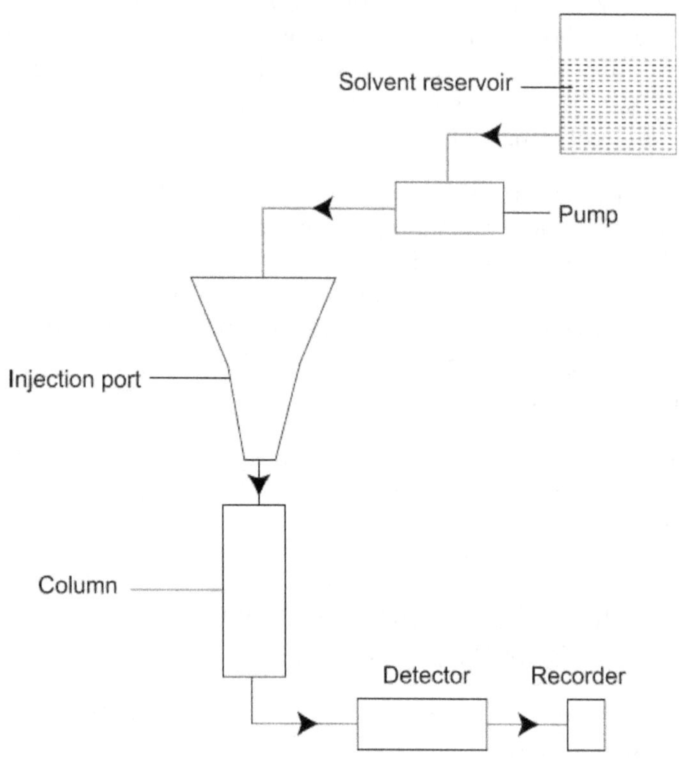

Figure 5.22. Components of a HPLC system (Schematic).

aluminum or copper. The column is straight with 20–50 cm length and 1–4 mm diameter. The supporting media in HPLC are rigid, solid column materials, which include three types namely micro porous supports, pellicular supports and bonded bases. In microporous supports, the particle has micropores having a diameter of 5–10 mm. The pellicular supports have a solid inert core surrounded by porous particles having a diameter of 40 mm. The bonded bases are the supports in which the particles are bonded to an inert support. Highly sensitive detectors such as UV photometers, spectrophotometers, refractive index monitors and fluorescence detectors are used to record the results.

Uses   HPLC is used for the separation of proteins, amino acids, polysaccharides, nucleic acids, plant pigments, pesticides, lipids and so on.

5.17 ELECTROPHORESIS

The movement of ions or molecules towards one of the electrodes under the influence of an electric field in a solution is called **electrophoresis.** For example, substances like sugar molecules are not charged and so they do not exhibit electrophoretic separation. On the other hand, proteins are colloidal in nature and undergo ionisation due to the possession of NH_2 and COOH groups so that they can be easily separated by electrophoresis.

5.17.1 Principle

Any substance suspended in water dissociates into charged particles. When these charged particles are subjected to an electric field, all positively charged ions would move towards the cathode or negative electrode and negatively charged ions towards the anode or positive electrode as shown in Figure 5.23. This is the basic principle involved in electrophoresis.

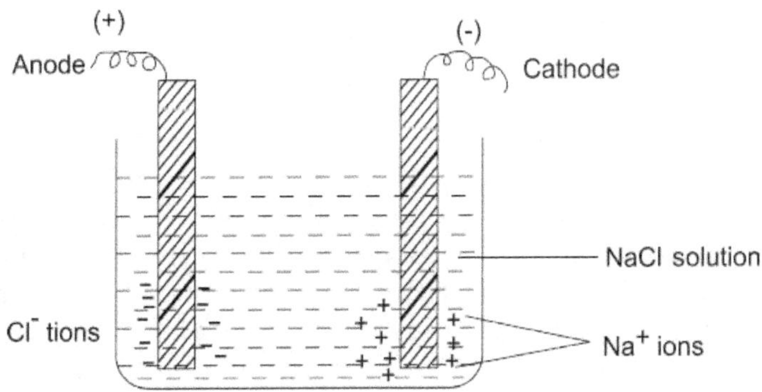

Figure 5.23. Movement of ions under an electric field.

5.17.2 Factors Affecting the Migration of Substances

Nature of charged particles   The electrophoretic mobility of substances depends on their size (molecular weight) and shape. Therefore, larger particles exhibit lesser electrophoretic mobility when compared to smaller particles. Smooth-surfaced molecules migrate faster than rough-surfaced substances.

Buffers   During electrophoretic mobility, a buffer solution provides the environment of charged particles, the required pH, electrolytic concentration, ionic strength and viscosity. The choice of the buffer solution depends on the nature of the sample to be separated. The commonly used buffers include citrate, phosphate, EDTA, acetate, Tris, barbitone, etc. Sometimes, a buffer could bind with sample components and affect their mobility. Such buffers are used to separate uncharged particles like carbohydrates. Thus borate buffer provides charges to sugars so that they can also be separated by electrophoresis.

Ionic strength   Substances with less ionic strength exhibit a faster separation whereas those with increased ionic strength show a slower separation.

pH   The degree of ionisation of substances depends on the pH so that their rate of separation is also affected by pH. An increase in pH causes ionisation of organic acids and the ionisation of organic bases increases in decreased pH.

Electric field   The current applied to the electrodes in a solution is carried mainly by the ions of the buffer solution even though some quantity is carried by the ions of the sample. The rate of movement of ions under potential gradient is called 'mobility'. The increased potential gradient increases the rate of movement of the particles. On the other hand, resistance plays an important role in the separation of particles. For instance, the electric current increases when the resistance is decreased and so the separation is faster. Moreover, the buffer ions carrying more charge than the ions of the sample would result in slower separation. Therefore, a constant current is to be maintained by using power packs during electrophoresis.

Electrodes   In electrophoretic studies, platinum, carbon or Ag/AgCl electrodes are used. Of these, platinum electrodes are mostly preferred. Though the use of carbon electrode is inexpensive, they are easily polarized and require frequent replacement. The silver electrodes are to be coated periodically. At each electrophoretic run, the polarity of the electrodes is to be reversed in order to prolong the life of electrodes and buffer solution.

5.17.3 Supporting Media in Electrophoresis

Various types of supporting media are used in electrophoretic separation of substances and are as follows.

Paper   Paper containing nearly 95% of cellulose with very low adsorption capacity can be used as a stabilizing medium in electrophoresis. However the adsorption of substances by the paper would affect the resolution of separation to some extent.

Gels   Gels are porous in nature and so the size of the pores in relation to size of the molecule determines the mobility of substances. As a result, the separation depends on the charge and size of the molecules when the gels are used as the supporting medium. For electrophoretic separation of components, the following types of gels are used.

Starch   Though the resolving power of starch gel is very high its pore size cannot be controlled. In addition, microorganisms contaminate it and it becomes opaque on staining. Therefore starch is not suitable for the separation of basic proteins.

Agar   Agar is soluble in aqueous buffer solutions and it forms a gel having a large pore size but without molecular sieving. Hence it is used to separate proteins and nucleic acids.

Polyacrylamide Polyacrylamide gel is prepared from components namely N, N'-methylene bisacrylamide (bis), ammonium persulphate and tetramethylene diamine (TEMED). It has low adsorption capacity but has no power of electrosmosis. Using differential concentration of the reagents can control the pore size of this gel.

Agarose–acrylamide It is a mixed form of gel obtained by mixing acrylamide with agarose. Here acrylamide provides sieve action while the agarose gives physical support to the gel. Therefore this is useful to separate compounds of very high molecular weight.

Other gels In addition to the above types of gels, substances like pectin, sephadex, gypsum, polyvinyl chloride, polyvinyl acetate, etc. are also used in electrophoresis, although only rarely.

TYPES OF ELECTROPHORESIS

Based on the mode of operation and separation, electrophoresis is of three types namely **boundary electrophoresis**, **zone electrophoresis** and **immunoelectrophoresis.** Among these methods, the boundary electrophoresis is seldom used whereas zone electrophoresis is more commonly put to use.

5.18 BOUNDARY ELECTROPHORESIS

This technique is the basis for all modern electrophoretic devices. The apparatus consists of a 'U' shaped cell provided with electrodes at the top position of the limbs as shown in the Figure 5.24.

The sample to be separated is dissolved in an appropriate buffer solution and the mixture is filled in the cell below the layer of buffer solution. The entire cell is kept in a temperature bath to have a constant temperature. The pH of the sample is adjusted in such a way that all the molecules of the sample

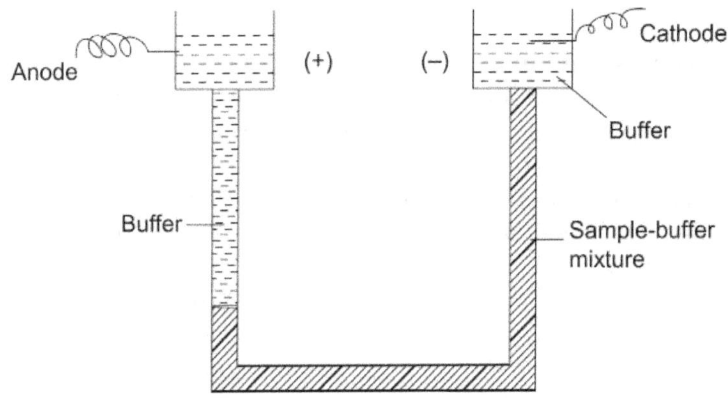

Figure 5.24. Schematic diagram of boundary electrophoretic apparatus.

possess a net negative charge. Under an electric field between the electrodes, all the molecules move towards the anode from the sample–buffer mixture to the region of buffer solution forming a boundary or front. As a result, the refractive index of the substance varies along the cell with a steep change at the boundary region. The change in refractive indices can be measured by optical devices which will exhibit the direction and the rate of migration of molecules in the sample. However all the fractions of the sample are not completely separated by this technique.

ZONE ELECTROPHORESIS

In this technique, the charged particles move as zones through a solid medium which may be a paper or powder-coated glass or silica gel. In other words, the separated fractions occur in the form of well-defined and permanent zones. Therefore, preservation of zones, interpretation of results and quantification of separated substances are easy. Zone electrophoresis includes two important types namely **paper electrophoresis** and **gel** or **disc electrophoresis.**

5.19 PAPER ELECTROPHORESIS

As represented in Figure 5.25, two compartments (beakers) containing buffer solution are interconnected by a supporting glass plate. The different buffer solutions used in paper electrophoresis are shown in Table 5.7. The level of buffer in the compartments is equalised. A suitable filter paper is selected and is made into strips having a width of 3–5 cm.

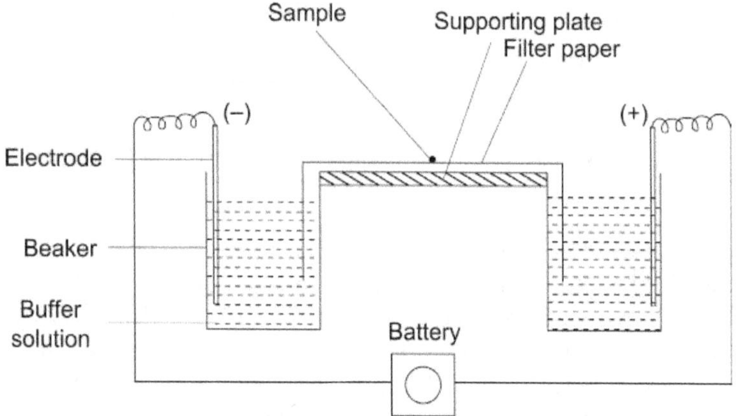

Figure 5.25. Set-up for paper electrophoresis.

Table 5.7. Buffers used in electrophoresis.

Substance	Buffer Solution	pH
Proteins	Barbiturate	8.6
	Borate	8.6
	Phosphate	7.4
Amino acids	Pthalate	5.9
Carbohydrates	Borate	10.0

At the centre of the strip, the origin line is marked with pencil in the form of a spot or a streak. The paper is now soaked in buffer solution and pressed between a blotting paper. The sample is applied on the spot with the help of a capillary tube or a micropipette without damaging the paper. Then the paper is placed on the supporting plate with each end dipped in the buffer solution of the opposite compartment. The electrodes of the compartments are connected to a power supply. The current is applied and the electrophoresis is run for 45 minutes at very low voltage (5–10 V/cm). When the separation is over, the paper is removed, dried and stained with locating reagents. Table 5.8 shows various locating reagents used in paper electrophoresis.

Table 5.8. Locating reagents used in paper electrophoresis.

Substance	Locating reagents
Proteins	Bromophenol blue
	Naphthalin black
	Azocarmine
Amino acids	Ninhydrin
Carbohydrates	Schiff-periodic acid reagent
Glycoproteins	Diphenylamine

After elution of fractions using a colorimeter, photoelectric scanning or a densitometer can make the quantification.

The paper electrophoresis used for the separation of substances at low voltage is called **low voltage electrophoresis**. This method is valuable in the separation of high molecular weight substances like proteins, carbohydrates and nucleic acids but it has only limited value in the separation of amino acids. When a high voltage current (200 V/cm for 10–100 minutes) is used in paper electrophoresis, it is called **high voltage electrophoresis** which is used for the separation of low molecular weight compounds and smaller ions like amino acids and inorganic substances. This technique requires a current of high voltage and longer duration (1000 V/cm and 18/24 hours). When the sample is allowed to flow continuously on the paper by some devices for the separation, then it is called **continuous flow electrophoresis**. The method is useful in the separation of amino acids and proteins.

5.20 CELLULOSE ACETATE ELECTROPHORESIS

Here cellulose acetate strips are used instead of paper to separate molecules with better resolution. While using cellulose acetate, the problem of adsorption is eliminated. Cellulose acetate is chemically pure, translucent and not very hydrophilic. The buffers for this technique are the same as those used in paper electrophoresis. But the solvents such as glacial acetic acid, cotton seed oil, liquid paraffin, etc. which make cellulose acetate transparent, are more preferred. Cellulose acetate electrophoresis is used to separate glycoproteins, lipoproteins and haemoglobin. In addition, immunoelectrophoresis uses cellulose acetate strips.

5.21 GEL ELECTROPHORESIS

In gel electrophoresis, gels are used as the medium instead of paper. When the gel is packed as a column, it is called **column zone electrophoresis**. If the gel is coated on glass plates, it is called **open block electrophoresis**. **Disc electrophoresis** is a column zone electrophoresis because gel is packed as a vertical cylindrical column in tubes.

5.22 POLYACRYLAMIDE GEL ELECTROPHORESIS (PAGE)

The apparatus consist of a buffer reservoir system and a power pack. The buffer reservoir system has an upper cathode compartment and lower anode compartment containing buffer solution. Glass tubes having 1.4 cm length and 0.5 cm diameter connect the two compartments. The compartments are provided with platinum electrodes. The schematic diagram of the apparatus is shown in Figure 5.26.

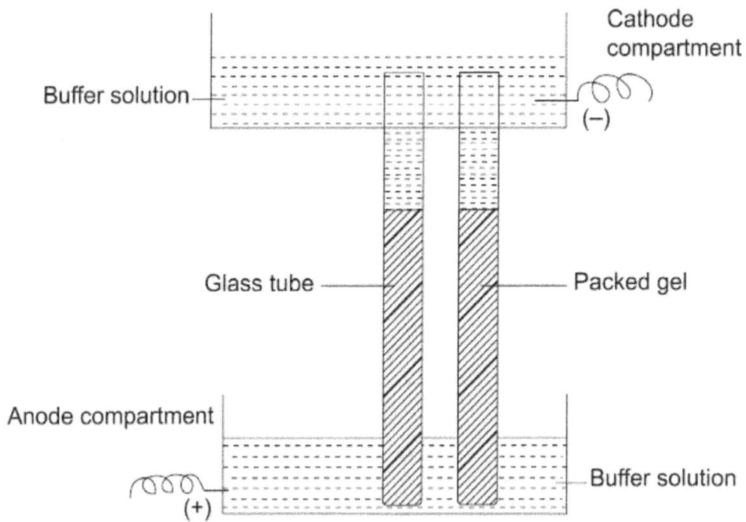

Figure 5.26. Schematic diagram of PAGE
(only two tubes are shown).

One end of each tube is tightly closed with a rubber cork and is filled with 1.1 ml of small pore solution (separation gel). Above this solution, 0.1 ml of double-distilled water is added and allowed to polymerize for 30 minutes. After the removal of water layer, 0.2 ml of large pore solution (spacer gel) is overlaid followed by 0.1 ml of distilled water for polymerization (15 minutes). Then the water layer is removed and 0.05 ml of the sample is added followed by 0.001 ml of marker dye. The remaining space of the tube is filled with Tris buffer. The preparation of reagents and buffer is shown in Table 5.9. After removing the rubber corks, the tubes are attached to the compartments in such a way that one end of a tube is attached to the anode compartment and the other end to the cathode compartment. Then the compartments are filled

Table 5.9. Preparation of reagents and buffer for PAGE.

Reagents		Quantity / Ratio
Solution A	1N HCl	12.00 ml
	Tris	9.15 gm
	TEMED	0.115 ml
	Water	50.00 ml
Solution B	1N HCl	12.00 ml
	Tris	1.40 ml
	TEMED	0.115 ml
Solution C	Acrylamide	15.00 gm
	Bisacrylamide	0.40 gm
	Water	50.00 ml
Solution D	Acrylamide	2.50 gm
	Bisacrylamide	0.625 gm
	Water	25.00 ml
Solution E	Riboflavin	4.00 mg
	Water	100.00 ml
Small Pore solution (pH 8.9)	Solution A: Solution B: Solution E: Water	1:1:1:1
Large pore solution (pH 6.9)	Solution B: Solution D: Solution E: Water	1:2:1:4
Tris buffer (pH 8.3)	Glycine	28.00 gm
	Tris	6.00 gm
	Water	100.00 ml

with Tris buffer (200 ml) and current is applied. When the marker dye migrates to the required position in the tube, the electrophoretic run is stopped and the tubes are removed from the compartments. The gels are immediately removed from the tubes with the help of a syringe needle and each gel is stored in 70% acetic acid contained in a test tube. The gels are stained with appropriate dyes and the sample is quantified spectrophotometrically or densitometrically.

5.23 SDS–POLYACRYLAMIDE ELECTROPHORESIS (SDS–PAGE)

This method is used for the study of the subunits of oligomeric proteins. The oligomeric proteins are formed by more than one polypeptide chain and are stabilized by hydrogen and disulphide bonds and also by hydrophobic interactions. Therefore these proteins move as a single unit during gel electrophoresis. In order to separate the subunits of these proteins, the solubilizing agents called solubilisers denature their structure. Urea, sodium dodecyl sulphate and b mercaptoethanol are mostly used as solubilisers. In concentrated urea, the hydrogen bonds readily dissociate. Sodium dodecyl sulphate (SDS), an anionic detergent, disrupts hydrophobic interactions and provides negative charge to the denatured polypeptides. The disulphide bonds are broken by mercaptoethanol.

In SDS–PAGE, SDS incorporated to the gel is used to separate individual polypeptide chains from oligometric proteins. The gels which have the property of molecular sieving exhibit a linear relationship between the electrophoretic mobility of protein, incorporated to SDS and the molecular weight of proteins. Therefore molecular weight of proteins can also be determined by this method.

5.24 SLAB GEL ELECTROPHORESIS

In this type, the polyacrylamide gel is allowed to polymerize into a thin slab between two glass plates. At one end of the gel, placing a comb into the gel prior to its polymerisation makes wells for spotting samples. After polymerisation of the gel, the comb is removed and a number of sample wells are got in the form of casts. As a result, a number of samples can be analysed and compared at a time.

5.25 TWO-DIMENSIONAL GEL ELECTROPHORESIS

It is a powerful tool and is designed by combining the resolving power of isoelectric focussing with SDS–PAGE. As the molecular weight and isoelectric point of a macromolecule are not related with each other, this technique makes use of these two properties to separate the molecules with great resolution power. By this method, a mixture of large number of proteins can be resolved into individual fractions.

In this technique, a protein mixture is subjected to isoelectric focussing on gel in a capillary tube. In contrast to the conventional electrophoresis in which the pH between anode and cathode is constant, in isoelectric focussing, a pH gradient is maintained with gradual increase from anode to cathode. When the protein sample is introduced into the system at pH below its isoelectric point, it has a net positive charge and will migrate towards the cathode (proteins possess a net positive charge in an acid medium). Therefore, the protein molecule moves to a medium of higher pH due to the presence of pH gradient. High pH of the medium influences the ionization and charge of the molecule, so the number of positively charged particles decreases. Thus at a particular pH (isoelectric point), the net charge of the protein becomes zero and will not move further. At this point, the gel is removed from the capillary tube, kept on the sample wells of a slab gel and subjected to SDS-PAGE. Now, the proteins are separated according to their molecular weight.

Uses   Gel electrophoresis is widely used to separate and isolate a large number of macromolecules. In molecular biology, this is a versatile technique used for determining sequences of DNA molecules, in verifying nature of nucleic acids and in restriction mapping of DNA.

5.26 IMMUNOELECTROPHORESIS

It is a combined technique of electrophoresis and specific immune reaction. A glass slide is coated with agar or agarose gel and rectangular and round holes are made in the gel as shown in the Figure 5.27. The round wells are filled with two different antigens (1–100 mg) and the gel plate is subjected to electrophoresis at low voltage for 1–2 hours. Then the plate is removed from the electrophoretic chamber and the rectangular well is filled with appropriate antiserum. The plate is incubated overnight at room temperature in a moist chamber, dried in an incubator at 37°C, fixed in 2% acetic acid and stained with protein stain. Precipitin bands as lines result at the regions where the molecules have separated by electrophoresis.

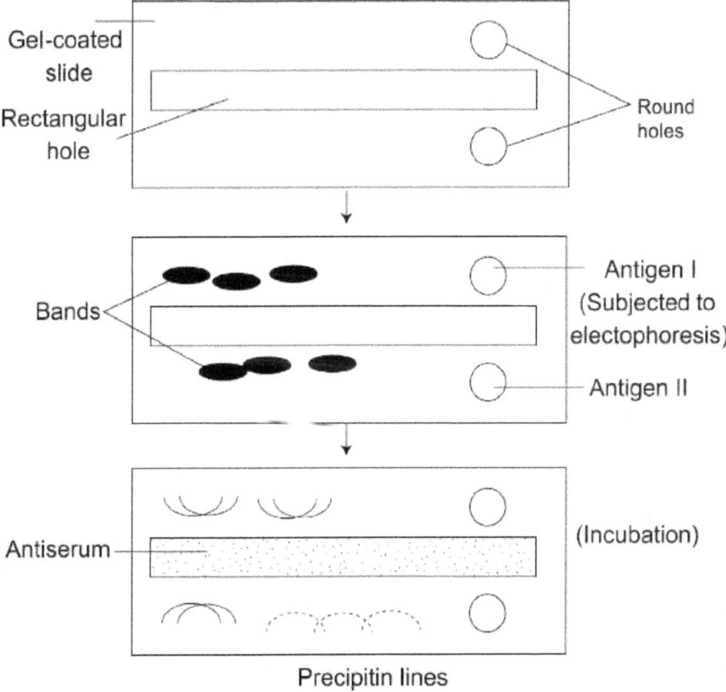

Figure 5.27. Steps in immunoelectrophoresis.

This technique is useful for identifying antigens in serum and to test purity of antigens. Thus it is used to identify bone marrow cancer and to study the gamma globulin in degenerative diseases of nerve fibres.

REVIEW QUESTIONS

1. Describe the pH scale.

2. How can you determine the pH of a sample using indicator?

3. What is the principle underlined in pH meter?

4. Write an essay on electrode system in pH meter.

5. Describe the structure of a hydrogen electrode.

6. Describe the structure of a calomel reference electrode.

7. Describe the structure of a typical glass electrode.

8. Describe the structure of an oxygen electrode.

9. Describe the structure of a blood gas electrode.

10. Describe the structure of a clark electrode.

11. Describe the structure of a carbon dioxide electrode.

12. What are the factors, which affect the measurement of pH in samples?

13. Write an essay on buffer systems in the biology of organisms.

14. Explain the bicarbonate buffer system in man.

15. Explain the phosphate buffer system in man.

16. Explain the protein and amino acid buffer system in man.

17. Ionization of biomolecules is pH-dependent. How?

18. Define centrifuge and centrifugation.

19. What is the basic principle of centrifugation?

20. Describe the structure of an ultracentrifuge.

21. Describe the structure of a preparative ultracentrifuge.

22. How is differential centrifugation be useful in cell fractionation?

23. What is the principle of density gradient centrifugation?

24. What is the principle of rate-zonal centrifugation?

25. What is the principle of isopycnic centrifugation?

26. Describe the structure of an analytical ultracentrifuge.

27. How can you determine molecular weight of substances using ultracentrifuge.

28. What is the principle of chromatography?

29. Explain various interactions that occur during chromatographic separation of substances.

30. Describe the technique of paper chromatography.

31. Describe the technique of two-dimensional paper chromatography.

32. Describe the technique of thin layer chromatography.

33. Describe the technique of column chromatography..

34. Describe the technique of gel permeation chromatography.

35. Describe the technique of ion exchange chromatography.

36. Describe the technique of affinity chromatography.

37. Describe the technique of gas-liquid chromatography.

38. Describe the technique of high performance liquid chromatography.

39. What are locating agents and how they are useful?

40. What is the principle of electrophoresis?

41. Give an account of various factors that affect the migration of substances during electrophoresis?

42. Explain various supporting media used in electrophoresis.

43. Describe the technique of boundary electrophoresis.

44. Describe the technique of paper electrophoresis.

45. Describe the technique of cellulose acetate electrophoresis.

46. Describe the technique of polyacrylamide gel electrophoresis.

47. Describe the technique of SDD–PAGE.

48. Describe the technique of slab gel electrophoresis.

49. Describe the technique of two- dimensional electrophoresis.

50. Describe the technique of immunoelectrophoresis.

ELECTROMAGNETIC RADIATION AND SPECTROSCOPY IN BIOLOGICAL STUDIES

6.1 ELECTROMAGNETIC RADIATION

Electromagnetic radiation is a form of energy transmitted through space with enormous velocity without requiring any supporting media. It has both wave and particle properties. In other words, it has an alternative field of electrical and magnetic force, thus possessing an electric component and a magnetic component. These two are mutually perpendicular to each other and to the direction of propagation as shown in the Figure 6.1. The best example for electromagnetic radiation is light.

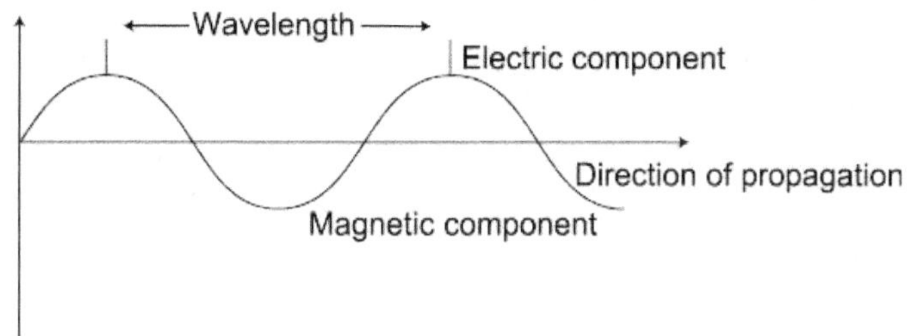

Figure 6.1. Electromagnetic radiation moving along an axis.

6.1.1 Wave Properties of Eelectromagnetic Radiation

The distance between two successive maxima of an electromagnetic wave is called wavelength (λ). If a radiation beam has only one discrete **wavelength**, then it is **monochromatic beam** and if it has several wavelengths, then it is **heterochromatic** or **polychromatic beam**. The total number of wavelength units, which pass through a given point at unit time, is called **frequency** (v). The wave number ($\bar{v}$) represents the number of waves/cm in vacuum. The velocity (V) of an electromagnetic radiation is the product of its wavelength and frequency.

Velocity (V) = wavelength (λ) x Frequency (v)

The velocity also depends on the medium through which the electromagnetic radiation passes. The relationship between frequency, wave number and velocity can be shown by the formula,

$$V = C\bar{v}$$

where,

C is the velocity of light through vacuum.

6.1.2 Particle Properties

Electromagnetic radiation exhibits refraction and reflection and possesses a stream of discrete particles of energy called **photons** or **quanta**. Photons travel in the direction of propagation of radiation. The energy of a photon is proportional to the frequency of radiation.

6.1.3 Electromagnetic Spectrum

The entire range over which electromagnetic radiation exists is known as **electromagnetic spectrum** which includes a wide range of wavelengths. Figure 6.2 shows a schematic diagram of different regions of electromagnetic radiation and the energy changes induced in matter. The radiation is represented in terms of its wavelength, frequency or the wave number. The wavelength is expressed in nanometers (10^{-9} m), or Angstrom (10^{-10} m). The term mμ (millimicron) is equivalent to nm. The γ rays are the shortest waves emitted by the atomic nuclei. The X-rays are emitted or absorbed by the movement of electrons. The ultraviolet region of the spectrum forms the beginning for chemical analysis. In the visible region of the spectrum, the human eye can identify and correlate the wavelengths.

The infrared region of the electromagnetic spectrum is associated with the changes in the vibration of molecules. The

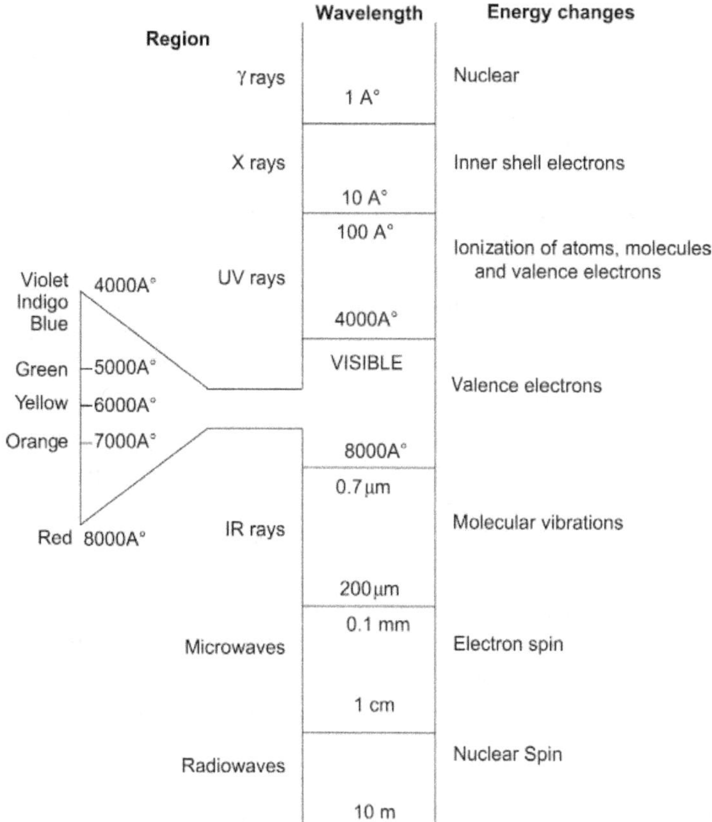

Figure 6.2. Schematic diagram of electromagnetic radiation and energy changes.

microwave region corresponds to the changes due to the rotation of molecules. The radio frequency region involves the energy change due to the reversal of rotation of nucleus or electron.

SPECTROSCOPY

The interaction between electromagnetic radiation and matter forms the basis for spectroscopy. When the electromagnetic radiation having a particular frequency is passed through a sample, the frequency which is absorbed or emitted by the molecules of the sample, is determined at molecular energy level by analysing the intensity of frequency emerging out of the sample. The instrument called **spectrometer** measures the change in intensity of electromagnetic radiation. The results are interpreted in terms of stereochemistry of the molecule (three-dimensional spatial arrangements of atoms in a molecule).

The spectrum obtained by the interaction in which a sample itself emits radiation is called **emission spectrum.** The spectrum obtained by the interaction in which the sample absorbs radiation is called **absorption spectrum.** When atoms absorb electromagnetic radiation, their electronic energy is increased resulting in a spectrum in the form of spectral lines of varying wavelength. This is **atomic absorption**. On absorption of electromagnetic radiation, the molecules become excited. When the molecules return to their original ground state, they emit molecular spectrum in the form of bands, each line of which is composed of a large number of very fine lines. This process is called **molecular absorption.** The energy associated with the uniform motion of a molecule is called **translational energy.** A molecule also possesses some forms of internal energy including **rotational energy,** which is due to the overall rotation of molecules, **vibrational energy** which is associated with the oscillation of atoms of the molecule, and **electronic energy** which is due to the movement of electrons. In a molecule, the levels of rotational energy are closely spaced so that very little energy is required for rotational transitions. This occurs in infrared and microwave regions of the electromagnetic spectrum and is useful in the study of molecular structure. The vibration energy levels are spaced apart and occur in infrared region of the spectrum. When molecules absorb photons of electromagnetic radiation in the visible and ultraviolet regions, the electronic energy is increased causing rotational and vibrational changes. As a result, the molecular spectrum exhibits a band of wavelengths.

6.2 ULTRAVIOLET AND VISIBLE SPECTROSCOPY

When electrons in an atom are excited, radiations are absorbed resulting in the absorption spectrum of the atom. Similarly, when the electrons of a molecule are excited, the electronic spectrum originates due to a change in the arrangement of molecular electrons. The electronic spectra of molecules occur in the visible and ultraviolet regions of electromagnetic radiation. A small change in electronic energy results in a large change in vibrational energy, which in turn results in a change in rotational energy of the molecule. In other words, the molecular electronic spectra involve change in electronic, vibrational and rotational energy of the molecule. The **UV absorption spectra** are due to the transition state of an electron or electrons within a molecule or due to an ion, which occurs from a lower to higher electronic energy level. The **UV emission spectra** are due to the reversion of transition.

The ultraviolet spectrometer consists of the following components:

Radiation source   The source of radiation may be hydrogen discharge lamps, deuterium lamps, xenon discharge lamps or mercury arcs. In all these lamps, the electrons are passed through a gas so that electronic vibrational and rotational excitations are produced in gas molecules.

Monochromators   These are dispersion elements, which may be prisms or gratings made up of glass, quartz or fused silica. The monochromator disperses radiation based on the wavelength.

Detectors   Three major types of detectors are used namely photovoltaic cell, photocell and photomultiplier tube.

Photovoltaic cell   It consists of a semiconductor (selenium) plate tightly fixed on a strong iron base (Figure 6.3). The plate is coated with a thin layer of silver or gold so that it acts as an electrode. When radiation falls on the surface of the plate, electrons are produced at the selenium–silver interface and are accumulated on its silver surface. This produces electrical voltage, which is proportional to the intensity of the incident radiation.

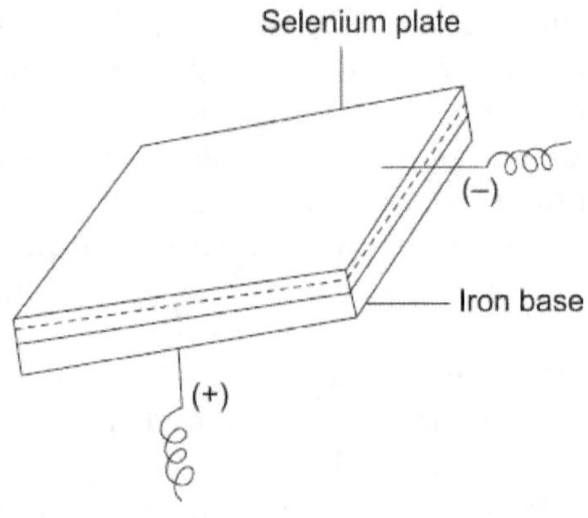

Figure 6.3. A photovoltaic cell.

Photocell   It has a light-sensitive metal cathode and anode placed inside an evacuated glass tube. The inner surface of the tube is coated with light-sensitive substance (Figure 6.4). When

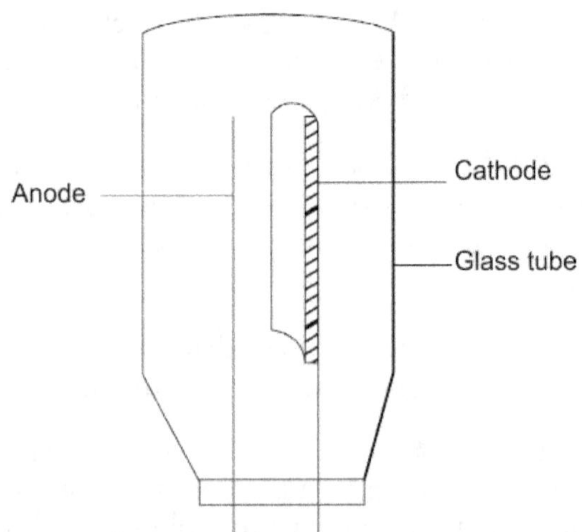

Figure 6.4. A photocell.

light falls on the surface coating, electrons are emitted and are collected by the anode. The resulting current is equal to the intensity of radiation.

Photomultiplier tube It has an evacuated tube containing one metallic cathode and anode with many dynodes (Figure 6.5). When radiation falls on the cathode, electrons are emitted and attracted by successive dynodes with the emission of more electrons at each dynode. As a result, a large number of electrons are produced and are proportional to the intensity of the falling radiation.

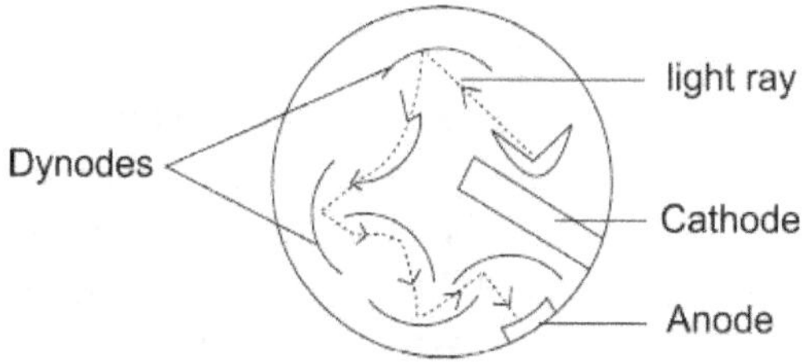

Figure 6.5. A photomultiplier tube (sectional view).

Recorder It receives the signal from the detectors and records the results.

Uses Visible and UV spectroscopy are useful in determining the concentration of biological substances. As the bases of the nucleic acids absorb UV rays, UV spectroscopy is used to study denaturation of DNA double helix. It is also employed to characterize aromatic compounds, to detect impurities in organic substances, to determine molecular weight of many compounds, to know the dissociation constant of acids and bases and to study kinetics of reactions.

6.3 INFRARED SPECTROSCOPY

The basic principle of Infrared spectroscopy is selective absorption of chemical substances in the infrared region producing infrared absorption spectrum in the from of bands. The spectrum includes a wide range of wavelengths, the intensities of which can be measured as transmittance or absorbance. The IR spectrometer consists of the following components.

Radiation source   The source of radiant energy may be an incandescent lamp, Nernst Glower, or Globar source or mercury arc.

Monochromator   In order to get the radiation source at desired frequencies a prism or grating is used as monochromator.

Sample and sample cells   As this technique is used to characterize solids, liquids and gases, these samples must be transparent to IR radiation. Therefore, the solid sample is prepared either by dissolving it in a suitable solvent or by grinding it with mineral oil into a paste. The liquid sample is directly taken in cells made up of NaCl or KBr. The gas sample is taken in a special cell, the end walls of which are made up of NaCl.

Detectors   Thermal detectors such as bolometers, thermocouples, thermisters, golay cell, photoconductivity cell, etc. are widely used along with an amplifier in order to identify even the low signals.

The IR spectrometer includes two types namely, **single beam IR spectrometer** and **double beam IR spectrometer**, the optic principle of which are shown in Figures 6.6 and 6.7.

In single beam IR spectrometer, the radiation from the source is allowed to pass through the sample and then into a mirror through a prism. The mirror and prism select the desired wavelength, which is amplified and detected. By knowing the intensity of original radiation, the absorbed radiation can be measured. However, the spectrum obtained by this instrument is deformed due to variations in intensity of the emission. The spectrum includes bands of solvents also when the sample is analysed in solution. These disadvantages are overcome in double-beam IR spectrometer.

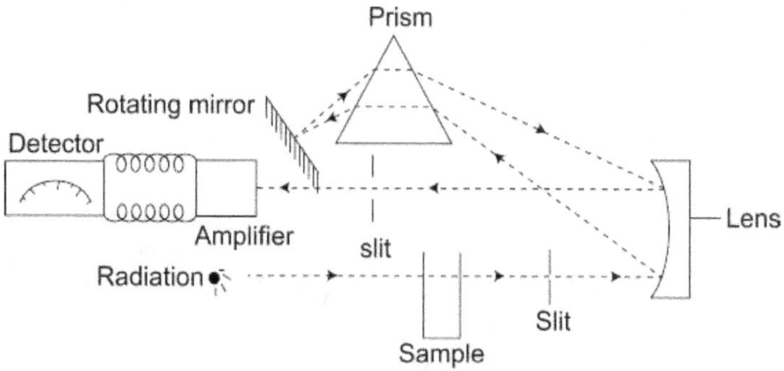

Figure 6.6. Optical system of a single-beam IR spectrometer.

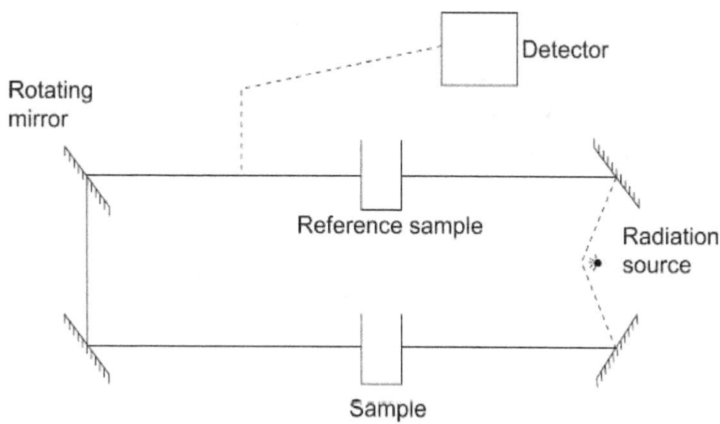

Figure 6.7. Optical system of a double-beam IR spectrometer.

In double beam IR spectrometer, the radiation from the source is split into identical beams, one of which is passed through the sample and another one through reference sample. Then the two beams are recombined and passed to the detector.

Uses IR spectroscopy is useful in identifying functional groups of organic compounds. It is used to determine steroids and hormones, to measure the rate of reactions, to study molecular conformations of biological molecules and to know the interactions between molecules.

6.4 FLUORESCENT SPECTROSCOPY

Fluorescent spectroscopy is based on the absorption phenomenon of electromagnetic radiation. When a molecule absorbs electromagnetic radiation, a part of energy is dissipated as heat and the remaining part of the energy is radiated with lesser frequency than the incident radiation. This is called **fluorescence.** In other words, the absorption of light by a molecule with a specific wavelength and its emission at a different wavelength is called fluorescence. Fluorescent spectra are always the band spectra, which are independent of the wavelength of the absorbed radiation.

The optical system of fluorescent spectroscopy is explained in the Figure 6.8. Here, the source of radiant energy is supplied by a mercury lamp or by a xenon arc.

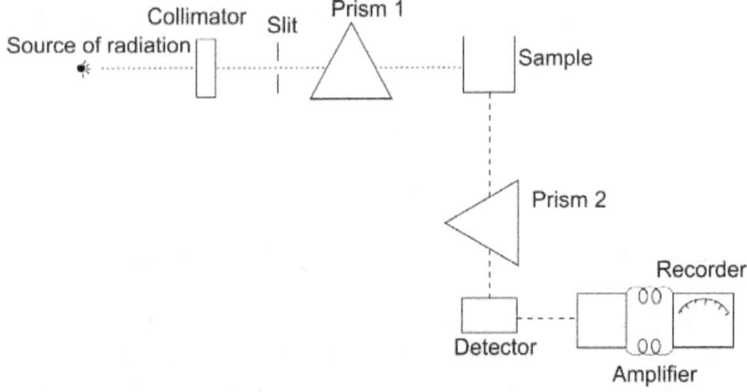

Figure 6.8. Optical system of a fluorescent spectrometer.

The radiant energy is allowed to pass through a collimator to get a pencil beam of radiation. Two prisms are used as monochromators of which one is placed in front of the sample to get suitable wavelengths and the one behind the sample to get fluorescent spectrum of the sample. A detector, which is usually a photomultiplier, an amplifier and a recorder are also provided.

Fluorescent spectroscopy is mainly used for studying protein structure as well as for qualitative and quantitative analysis of compounds.

ATOMIC SPECTROSCOPY

When a polychromatic light beam is passed through a prism, its constituent colours are separated in the form of its spectrum. The spectrum thus formed is of two main types namely **emission spectrum** and **absorption spectrum.** The emission spectrum is obtained by passing the light through a prism and can be examined directly by a spectroscope. When a light is passed through an absorbing substance, some colours of the spectrum appear as dark bands under a spectroscope. Such a spectrum is called absorption spectrum.

6.5 ATOMIC ABSORPTION SPECTROSCOPY

When a metallic solution is introduced into a flame, the metallic vapour is formed in which some metal atoms emit radiations characteristics of the metal. But a large percentage of atoms remain as non-emitting ground state. These ground state atoms are capable of receiving light of specific wavelength. When such a light is passed through the flame having the metallic atoms, a part of light is adsorbed and is proportional to the density of the atoms.

The important components in an atomic absorption spectroscope are given in the Figure 6.9. It consists of the following components:

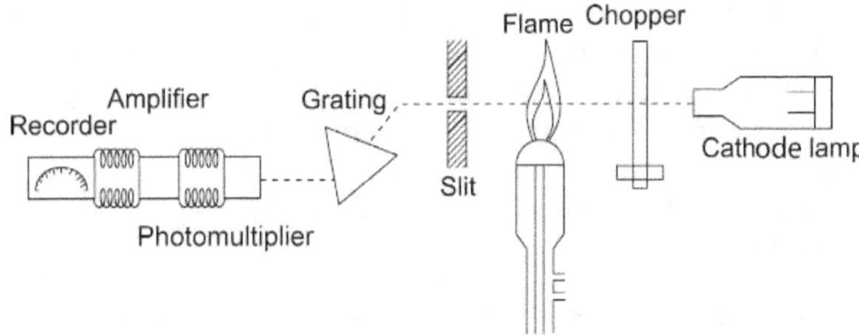

Figure 6.9. Optical system of an atomic absorption spectrometer.

1. *Radiation source*   A hollow cathode lamp is used as the radiation source in this instrument. It produces the spectrum of the metal, which is used as cathode. Gaseous discharge lamps or arc lamps containing an inert gas at low pressure and a metal and metal salt are also used.

2. *Chopper or Rotating wheel*   It is placed in between the cathode lamp and the flame for breaking the steady light into an intermittent one so that the interference of light from the flame is avoided.

3. *Burner*   The burner produces the flame, which converts the liquid sample into atomic vapour. The fuel, oxidizing gas and the sample are passed into the burner so that the flame makes the liquid sample into atoms. Prior to the entry of the sample liquid into the burner, it is converted into small droplets by passing a gas at high velocity (**pneumatic nebulization**).

4. *Monochromator*   Prism or gratings are most commonly used as monochromators to select a particular wavelength of light.

5. *Detector and amplifier*   The light from the monochromator is received by a photomultiplier tube and is amplified by the amplifier.

6. *Recorder* Even though digital recorders are used in some instruments, mostly chart recorders are employed in this technique.

6.6 DOUBLE BEAM ATOMIC ABSORPTION SPECTROSCOPY

In this method, one part of the chopped light beam from the lamp is passed through the flame and the other part is made to bypass the flame. Then the two beams are recombined and passed through the monochromator. Therefore, the drifting effect of the lamp and the change in the sensitivity of the detector are eliminated.

Uses This technique is mainly useful in determining metallic elements such as, Ca, Mg, Na, K, etc. in biological substances. As this method requires no sample preparation, both aqueous as well as non-aqueous solutions can be studied.

6.7 ATOMIC EMISSION SPECTROSCOPY

The important components of an emission spectrometer are shown in the Figure 6.10. It consists of the following parts:

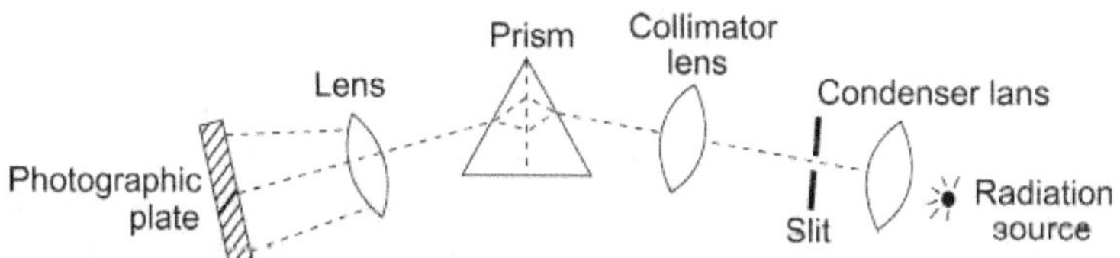

Figure 6.10. Optical system of an emission spectrometer (with prism as monochromator).

1. *Radiation source* The radiation source in emission spectroscopy includes flames, direct current arc or alternating current spark arcs which are capable of vaporizing and dissociating the sample into atoms of high energy level above the ground state.

2. *Sample and sample holder* In this method, both solids and liquids can be analysed. If the sample is a solid and a good conductor with the capacity to withstand high temperature, it is made into an electrode (self electrode). For other solid substances, the sample is powdered, mixed with graphite powder and placed to the lower graphite electrode. When electric current is passed through the upper side electrode, the sample gets vaporized. For liquid samples, the sample is directly taken in the sample holder which is provided with mechanism to discharge the sample into the counter electrode.

3. *Monochromator* In this spectroscopy, prisms or gratings are the monochromators. The prism is a mostly two-half prism made of quartz or fused silica. The first half of the prism splits the light into two beams and the second half recombines the split beams into a single beam.

4. *Detectors* Photomultiplier tubes and photographic plates are widely used for quantitative and qualitative analyses of samples respectively. The photomultiplier tubes are useful for reading the emission spectra directly. The spectra are recorded on a photographic plate and the intensity of the spectrum is measured in a densitometer.

Uses This technique is used to analyse various elements in the tissues of animals to know the changes in the concentration of trace elements during ageing process. It is also used to determine the deficiency of elements in plants and soils.

6.8 MASS SPECTROSCOPY

Mass spectroscopy helps to separate atoms or molecules in a compound on the basis of differences in their masses. When a compound is bombarded with a beam of electrons its ionic molecules are produced. These molecules separate according to their masses producing a spectrum called **mass spectrum.**

Diagrammatic structure of a typical mass spectrometer is shown in the Figure 6.11. It consists of the following components:

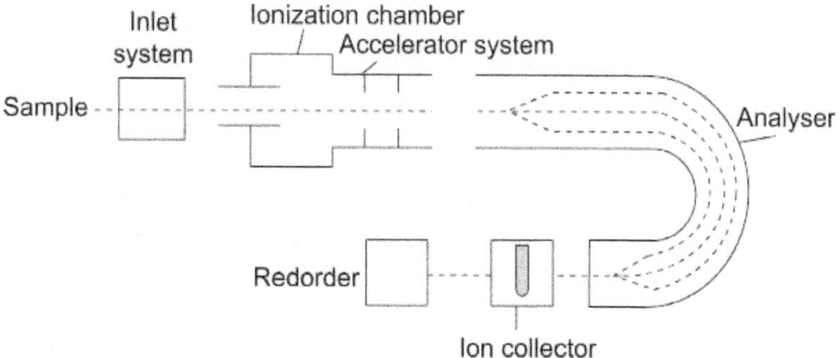

Figure 6.11. Diagrammatic structure of a mass spectrometer.

1. *Inlet system*   As the instrument requires a vaporized sample, the sample is converted into vapour in the inlet system by heating. The volatile sample is heated in a flask and introduced into the ionization chamber. If the sample is a gas, it is transferred from a gas bulb through melting and expansion reservoirs. Liquid sample is are injected by hypodermic needle through silicon rubber tube.

2. *Ionization chamber*   This provides the source of ions. Here a beam of electrons is passed through the molecules of the samples so that the molecules become ionized. There are a variety of methods for ionization of the sample and are as follows:

Electron impact ion source   This is used in most of the instruments in which an electrically heated filament produces electrons, which are accelerated by the anode. Thus, a beam of electrons intersects the sample molecule resulting in positively charged molecules in the form of a mass spectrum.

Knudsen cell   Here the sample in a crucible is vapourized through heating by radiation or by electron bombardment and is directed towards the analyser through a small hole.

Spark source ionization   By this method, the sample is made into two electrodes to which high voltage is applied. As a result, positive ions are produced and are evaporated.

Surface ionization   In this method, the solid sample is coated on a filament of ribbon and heated. Now the positive ions evaporate.

Chemical ionization   Here gas like methane is introduced along the sample under a beam of electrons so that the gas ionizes and interacts with the sample molecules to form positive ions.

3. *Accelerator system*   It includes electric fields, which receive the positive ions from the ionization chamber and accelerate the ions according to their masses. The accelerated ions are passed on to the analyser through a slit.

4. *Analyser or ion separator*   It is the most important component in the instrument as it separates the ions according to their masses. Based on the nature of analysers, various types of mass spectrometers are designed.

5. *Ion collector*   This includes photographic plates, electron multipliers or electrometers. A photographic plate provides greater resolution and sensitivity and so it is more advantageous.

6. *Vacuum system*   The entire instrument is kept in a vacuum-tight chamber.

Uses   Mass spectroscopy is used to determine the molecular weight of the substances, the quantity of isotopes in vaporizable elements, ionization potentials of molecules and bonding nature of macromolecules.

6.9 RAMAN SPECTROSCOPY

When a beam of monochromatic light ray is passed through a transparent substance (solid, liquid or gas), the incident light rays are mostly transmitted without alterations. However, some of the rays are scattered at different angles than their incident angles. This scattered light contains additional frequencies than that of the incident rays. This phenomenon is called **Raman effect**. The lines with modified wavelengths in Raman effect are called **Raman lines.** The lines with greater wavelength than that of the incident wavelength are called **Stoke's lines** and those with shorter wavelength are called anti-Stoke's lines. In Stoke's lines, the intensity is greater than the intensity of the corresponding anti-Stoke's lines. Raman frequencies are more or less similar to the frequencies of infrared regions of the spectrum.

The important components of Raman spectrometer are shown in the Figure 6.12. It consists of the following components:

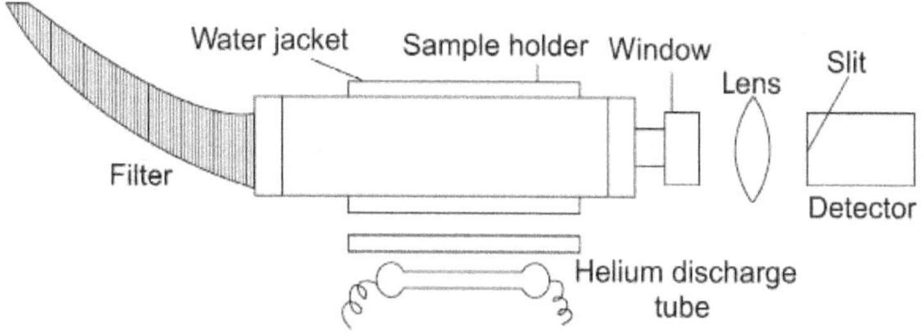

Figure 6.12. Diagrammatic structure of Raman spectrometer.

1.   *Source of light*   Raman effect requires a source of light of higher intensity. To get such an excitation, mercury lamps are used. While using mercury lamps, large volume of highly concentrated sample should be applied. Moreover, the sample should be clear, colourless and non-fluorescent. The above difficulties

are overcome by the use of helium-neon laser light. This source can be easily focused and results in better resolution and requires filter. Moreover, coloured solutions can also be studied using this lamp.

2. *Filters*   While using non-monochromatic source of light, filters are used to get monochromatic radiation. Filters made up of nickel oxide, glass or quartz glass and also coloured aqueous solutions like ferricyanide or iodine in carbon tetrachloride are used as monochromators.

3. *Sample holder*   The selection of sample holder depends on the intensity of light source and the nature of the sample. For example, gases require bigger sample holders than those required by the liquids. The sample holders is a glass tube, one end of which is drawn into a horn-like structure with a blackened outside. This end provides suitable background and the end is closed by a glass plate with a window in front. The entire tube is surrounded by a water jacket in which cold water is circulated to prevent overheating of the sample.

4. *Detector*   The scattered radiation is passed towards the slit present on the detector through a lens situated in front of the window of the sample holder. The detector contains a photomultiplier tube and a recorder. Raman lines are studied either through photographic plate or through automatic recording devices.

Uses   Raman spectroscopy is used to study the structure of gases, nature of bonding in molecules and the molecular structure of organic compounds. The technique is also used to study electrolytic dissociation, hydrolysis and change from crystalline state to amorphous state of molecules.

X-RAY SPECTROSCOPY

When accelerated electrons having high kinetic energy bombard a metal, the energy is lost inside the metal atom, which emits the energy as X-ray radiation. Thus, the impinging of high velocity electrons on a metal target produces X-rays. When the accelerating energy is increased, the bombarding electrons become highly energetic, producing X-rays with higher frequency and intensity. As a result, the X-rays are in the form of sharp peaks with a wavelength similar to metal targets. In this way, X-rays are electromagnetic waves produced by electrons of an atom and are scattered in all directions. This is because the electrons that get energy from the electromagnetic X-ray radiation, become excited and radiate the energy. X-rays interact with matter either by absorption or scattering or diffraction. When an X-ray beam is passed through matter, part of its energy is lost by scattering and a part by absorption. Scattering of X-ray radiation forms the basis for diffraction. In other words, the electrons in the atoms of the matter absorb energy from X-rays and become excited producing secondary radiation characteristic of the atoms.

X-ray spectroscopy includes three major types namely **X-ray absorption spectroscopy, X-ray diffraction spectroscopy** and **X-ray fluorescent spectroscopy.** All these instruments have more or less similar components but they differ only in their optical systems.

The structure of an X-ray spectroscope is shown schematically in the Figure 6.13. The important components of an X-ray spectrometer are as follows

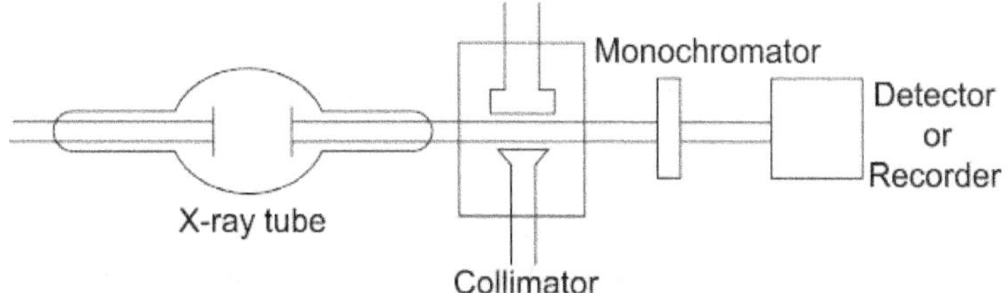

Figure 6.13. Diagrammatic structure of an X-ray spectrometer.

1. *X-ray source*   A variety of X-ray tubes are available to produce X-rays. A typical X-ray tube (Figure 6.14) consists of a cathode filament (tungsten filament) and an anode with target. The cathode emits electrons which are accelerated towards the target. The energy of electrons is now transferred to the metallic surface of the target which emits X-radiation.

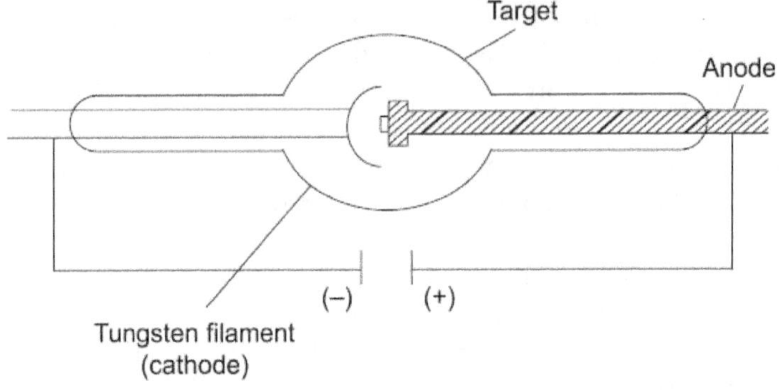

Figure 6.14. A typical X-ray tube.

2. *Collimator*   It consists of two closely packed metal plates with a small gap. This absorbs all X-radiations which are passed out in a narrow X-ray beam.

3. *Monochromator*   To get X-radiation of required wavelength, filters or crystal monochromators made of sodium chloride, lithium fluoride, quartz, etc. are used.

4. *Detectors*   The intensity of X-rays can be measured by photographic technique or by using radioactive counters.

6.10 X-RAY ABSORPTION SPECTROSCOPY

X-ray absorption spectrometer is designed specifically with a provision of ionization chamber and its diagrammatic structure is given in the figure 6.15. It consists of a tungsten X-ray tube, a motor-driven chopper to interrupt one half of the X-ray beam, an aluminium attenuator of variable thickness for making adjustment to balance absorption in the sample, a reference X-ray beam, a photomultiplier tube to receive X-ray beams, an amplifier and a detector. The results are interpreted through a calibration curve.

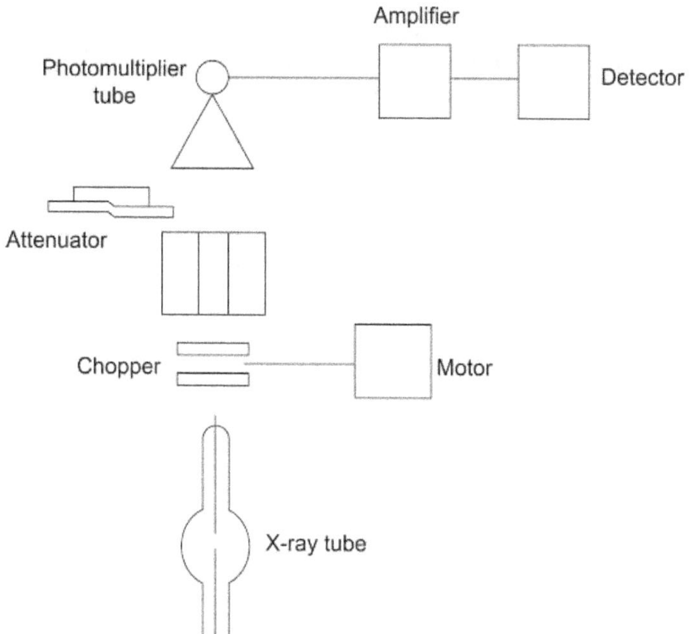

Figure 6.15. Diagrammatic structure of an X-ray absorption spectroscope.

Uses   This instrument is used to analyse substances qualitatively and quantitatively, to locate trace elements and to measure the volume of liquids in closed vessels.

6.11 X-RAY DIFFRACTION SPECTROSCOPY (X-RAY DIFFRACTION CRYSTALLOGRAPHY)

When an X-ray beam is passed through a substance, the electrons of its atoms emit electromagnetic radiation in all directions like that of the incident X-radiation. These scattered waves from the electrons are arranged in the form of a crystal lattice. The interference of these waves cause diffraction by the crystal plane. Thus each crystalline substance scatters the X-ray in the form of its own diffraction pattern according to its atomic and molecular structure. X-rays are electromagnetic waves having definite wavelengths and the atoms are found in regular three-dimensional structure in crystals. Therefore, X-radiations resolve the atoms found in a crystal and the atoms of the crystals now scatter the X-rays, thus producing a diffraction pattern in the form of spots. The diffraction pattern thus obtained is recorded on an X-ray film or an electronic detector to determine the structure of the molecules.

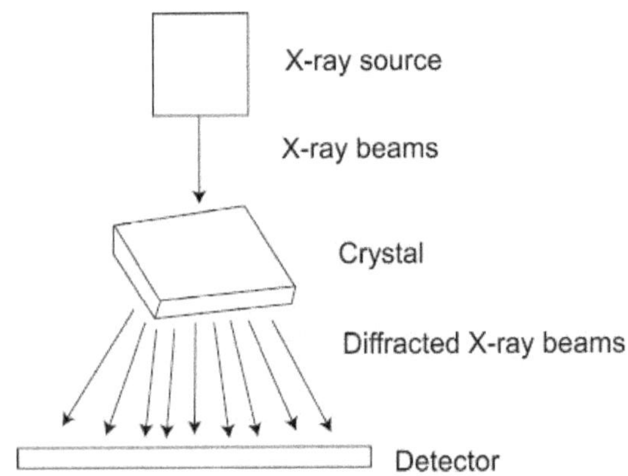

Figure 6.16. Schematic representation of X-ray crystallography.

The study of crystals constitutes **crystallography**. The crystals are three-dimensional space gratings and are used for diffraction of X-rays. When an X-ray beam having continuous range of wavelengths are allowed to pass through a crystal and the transmitted beam is made to fall on a photographic plate as in Figure 6.16, the photograph will show a central spot surrounded by a number of other weak spots in a definite pattern.

Thus the X-ray beam has been diffracted according to the geometrical pattern of the crystal in which the atoms are arranged in a regular three-dimensional lattice. The regular repetition of units in three dimensions is called **lattice** which may contain atoms arranged in many planes. Some planes will have more atoms than the others. When a beam of X-ray is passed vertically downwards on a molecule, there is a partial reflection of X-rays at each plane of atoms of the molecule. These reflected beams exhibit diffraction pattern.

Though biological structures are rarely crystalline in nature, X-ray crystallography is the only method to study the structure of these molecules. Here the biological molecule is crystallized, irradiated with a beam of X-rays and the diffraction pattern is analysed. The three-dimensional pattern obtained will indicate the position of each atom in a molecule.

The X-ray diffraction spectroscopy includes the following instrumentations:

6.11.1 Transmission Photographic Method

In this method, a tungsten X-ray tube produces a beam of X-rays which is allowed to pass through a pinhole collimator. Now a fine pencil of X-rays traverses the crystal to be studied. After crossing the crystals, the X-rays get diffracted and are recorded on a photographic plate as shown in the Figure 6.17.

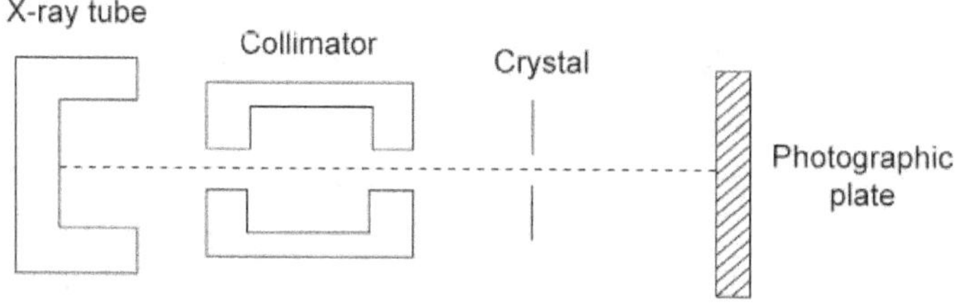

Figure 6.17. Schematic diagram of transmission photography.

6.11.2 X-ray Ionization Spectrometer

This instrument consists of an X-ray tube, which produces X-rays. X-rays are passed through two adjustable slits so that a thin beam of X-rays falls on the crystal sample. The reflected X-rays pass through a slit into the ionization chamber through a narrow aluminium window. One end of the ionization chamber is connected to the positive terminal of a battery and the negative terminal of the battery is connected to an electrometer to measure the ionization current. The meter measures the intensity of the reflected X-rays. The diagrammatic structure of ionization spectrometer is shown in the Figure 6.18.

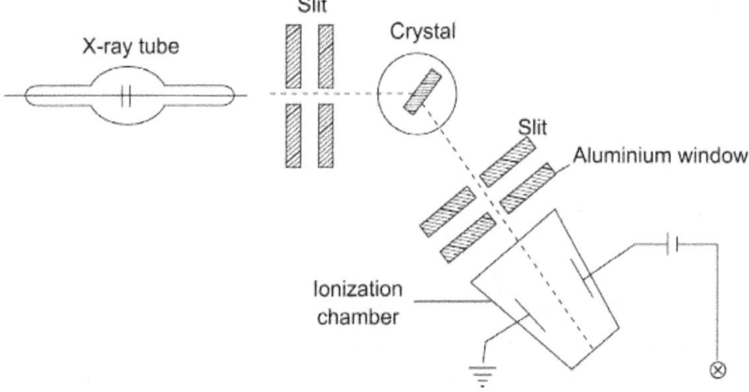

Figure 6.18. Diagrammatic structure of ionization spectrometer.

6.11.3 Rotating Crystal Diffraction Spectrometer

The important components of this instrument are diagrammatically represented in the Figure 6.19. Here the X-rays produced by the X-ray tube are passed through a filter and then through a collimator to form a fine beam of parallel X-rays. The X-ray beam is then allowed to fall on a crystal, which is placed on a rotatable shaft. The crystal is made to rotate slowly on a fixed axis in order to cause planes of reflecting positions. The diffraction pattern as spots can be photographed.

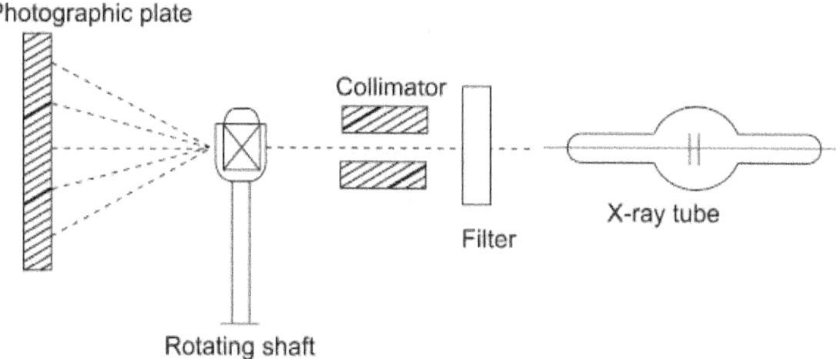

Figure 6.19. Diagrammatic structure of rotating diffraction spectrometer.

6.11.4 Powder Crystal Diffraction Spectrometer

Here the X-ray beam is passed through a filter on the powdered sample. The sample is placed vertically in the axis of a cylindrical camera. After traversing through the powder, the X-rays pass out of the camera through a small hole in the film and fall on a photographic plate in the form of lines of a particular pattern. The structure of this spectrometer is shown diagrammatically in the Figure 6.20.

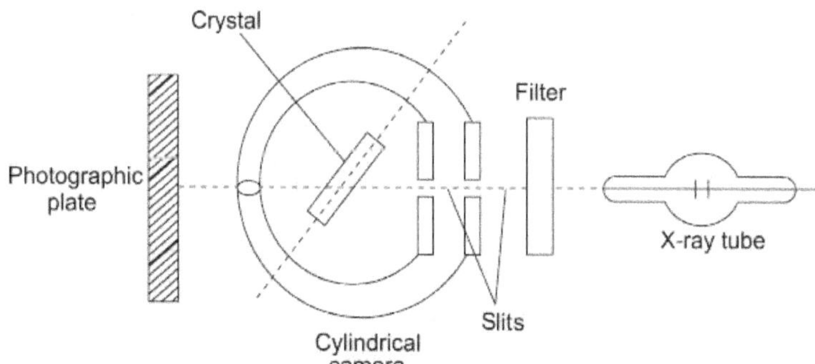

Figure 6.20. Diagrammatic structure of powder crystal diffraction spectrometer.

Uses   X-ray diffraction spectroscopy is generally used to differentiate 'cis' and 'trans' isomers and linkage isomers of a complex. The transmission photographic method is useful to study the orientation of fibres and the symmetry of a single crystal. Ionization spectroscopy is used to study crystalline structures of molecules, while rotating spectroscopy is used to determine the size of unit cells. Powder crystal spectroscopy is useful to study cubic crystals.

6.12 NUCLEAR MAGNETIC RESONANCE (NMR) SPECTROSCOPY

Radio waves are electromagnetic radiation with lowest energy and induced transition in magnetic level of the nuclei of atoms. Though the energy in radio frequency is very small, it affects the nuclear spin of atoms in a molecule so that the nucleus of an atom changes the direction of its spin on absorbing radiofrequency radiation. When a nucleus absorbs energy, it becomes excited and loses energy on return to unexcited state. The nucleus with alternate excited and unexcited conditions is said to be in a state of resonance. The absorption of radiation in the radiowave region of the electromagnetic spectrum by the nucleus in a magnetic field and its change from low energy state to high energy is called **nuclear resonance**. The energy absorbed by the nucleus is measured to determine the resonance frequency. When a nucleus absorbs energy, it produces a signal which is detected, amplified and recorded as a band of spectrum called **NMR spectrum.**

The important components of NMR spectrometer are diagrammatically represented in the Figure 6.21. The instrument consists of the following components.

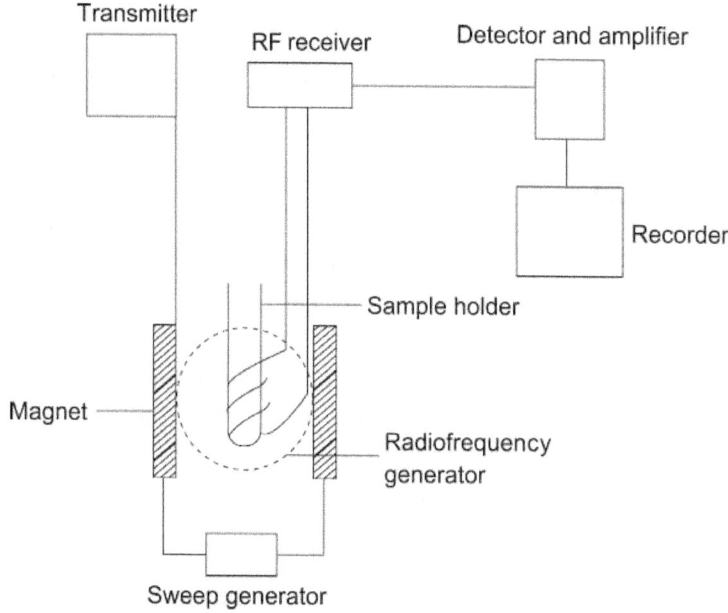

Figure 6.21. Diagrammatic structure of NMR spectrometer.

1. *Sample holder*   Glass tubes are generally used as they are chemically inert, durable and transparent to radiofrequency radiation.

2. *Magnet*   Permanent magnet or electromagnet is used. The magnet provides a homogenate magnetic field, the strength of which should be very high and constant.

3. *Sweep generator*   It has a pair of Helmholtz coils fixed in the pole spaces of the magnet. These coils produce magnetic field, which can be changed by varying current flow. Thus it brings varying applied magnetic field to bring nuclear resonance. The magnetic field from these coils is added on to the main field so that the sample is exposed to both magnetic fields. Therefore the sample becomes magnetized.

4. *Radiofrequency generator* A radiofrequency oscillator produces radiofrequency radiation. The coil of oscillator is wound around the sample holder in order to irradiate the sample with radiofrequency radiation.

5. *Radiofrequency receiver* In the magnetized sample, the frequency undergoes absorption and dispersion. The receiver determines the resonance frequency, which is directed to the detector.

6. *Detector and amplifier* A phase-sensitive detector separates the absorption signals from dispersion signals. As the absorption signal from the receiver is extremely weak, the amplifier amplifies it.

7. *Recorder* a chart recorder records the amplified signal.

Uses NMR spectrum of the sample is obtained by plotting the intensity of the signal against the strength of the magnetic field. By using this technique, the three-dimensional structure of small protein molecule can be determined. It is also useful to study the dynamics of protein molecules and small conformation changes, intracellular pH, membrane transport and concentration of metabolites in tissues and organs.

6.13 ELECTRON SPIN RESONANCE (ESR) SPECTROSCOPY

The absorption of radiations of a specific frequency by the electrons under a magnetic field and their change from low-energy state to high-energy state is called **electron spin resonance**. The two states of energy levels are produced by the interaction between the magnetic moment of an unpaired electron and applied magnetic field. Substances with unpaired electrons absorb radiation having the frequency in the microwave region. These electrons produce transitions between magnetic energy level resulting in **ESR spectrum,** which is measured by plotting the intensity of the signal obtained by the spectroscopy against the strength of magnetic field.

The electron spin resonance spectroscopy is more or less similar to nuclear magnetic resonance spectroscope but the sensitivity of ESR spectroscope is comparatively higher than NMR spectroscope and it requires only very little quantity of the sample. The important components of ESR spectrometer are shown in the Figure 6.22. It consists of the following parts.

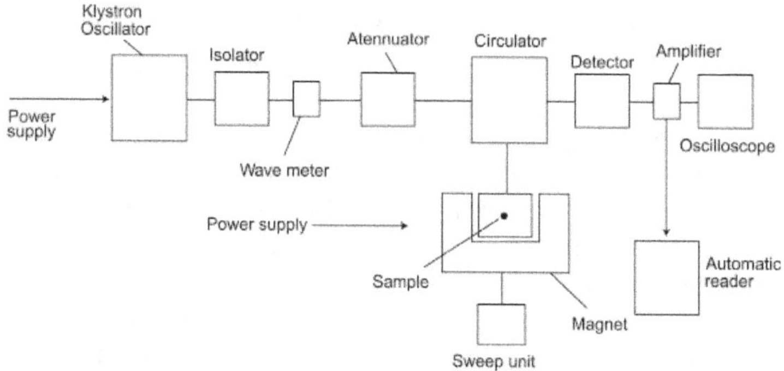

Figure 6.22. Diagrammatic structure of ESR spectrometer.

1. *Source of microwave radiation (klystron oscillator)* The klystron oscillator is a vacuum tube which produces microwave radiations on applying voltage. An isolator minimizes the vibrations in the frequency of microwaves. The frequency of microwaves is known from a wave meter. The microwaves are transmitted to a circulator by an attenuator.

2. *Circulator*　It is a hollow, rectangular tube formed by copper or brass with silver or gold plating to have a high-conducting flat surface. The microwave radiations are passed towards the sample from one end of the circulator and the end is attached to the detector.

3. *Sample cavity and magnet system*   Flat cells are used as sample holders. The sample holder is placed in a sample cavity in which maximum magnetic field is maintained throughout the sample. The sample cavity is placed in an electromagnet, which provides a uniform magnetic field. A pair of sweep coils are provided for varying the power supply.

4. *Detector*   A silicon crystal detector is mostly used. It converts the microwaves into electrical current, which is amplified in the form of a narrow band.

5. *Recorder*   The signal from the amplifier is recorded by an oscilloscope or by an automatic recorder.

Uses   The biological macromolecules do not resonate due to the absence of unpaired electrons. They are 'spin-labelled' and are used in ESR spectroscopy. In this way, the technique is used to determine the rate of enzyme reactions, to find substrate-binding region of enzymes, to study denaturation and folding of protein molecules, enzyme–ligand interactions and symmetry and orientation of protein molecules. The apparatus is also used to study free radicals and inorganic compounds.

REVIEW QUESTIONS

1. What are the important properties of electromagnetic radiation?

2. Describe the electromagnetic spectrum.

3. What are the different spectra obtained by the interaction of electromagnetic radiation with matter?

4. Describe the technique of ultraviolet and visible spectroscopy.

5. Describe the technique of infrared spectroscopy.

6. Describe the technique of fluorescent spectroscopy.

7. Describe the technique of atomic absorption spectroscopy.

8. Describe the technique of atomic emission spectroscopy.

9. Describe the technique of mass spectroscopy.

10. Describe the technique of Raman spectroscopy.

11. Describe the technique of X-ray spectroscopy.

12. Describe the technique of X-ray absorption spectroscopy.

13. Describe the technique of X-ray diffraction spectroscopy.

14. Describe the technique of nuclear magnetic resonance spectroscopy (NMR).

15. Describe the technique of electron spin resonance spectroscopy (ESR).

OTHER OPTICAL TECHNIQUES IN BIOLOGICAL STUDIES

7.1 OPTICAL ROTATORY DISPERSION (ORD) AND CIRCULAR DICHROISM (CD)

Most light as it occurs in nature is unpolarized and it consists of different wavelengths and vibrates in many different planes. If the magnetic field of light vibrates in a specific direction, then the light is said to be **linearly polarized.** If the linearly polarized light contains additional component of unpolarized light, it is **partially polarized.** If the electric field, instead of oscillating in a plane, proceeds in the form of a helix around the axis of propagation with a constant magnitude, then the light is referred to as **circularly polarized.** Here, the electric field completes one revolution within one wavelength. If the tip of the field rotates clockwise, then the light is called **right circularly polarized light** (RCPL). If it rotates anticlockwise, then it is **left circularly polarized light** (LCPL) (Figure 7.1). **Elliptical polarization** is the intermediate condition between circular and linear polarizations. In other words, linear and circular polarizations are two extremes of elliptical polarization. Light can be polarized by scattering, reflection, transmission, selective absorption or double refraction.

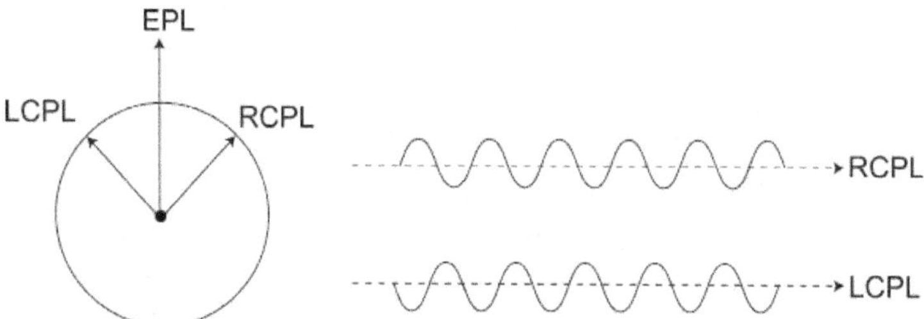

Figure 7.1. Rotation of circularly polarized light.

(CPL – Circularly polarized light; RCPL – Right circularly polarized light ; LCPL– Left circularly polarized light)

7.1.1 Optical Activity

When a beam of linearly polarized light passes through a substance, the light remains linearly polarized but its plane of oscillation rotates. i.e., its plane of vibration becomes changed in orientation. Such substances are called **optically active** and the effect is called **optical activity** or **optical rotation.** When looked towards the source, the rotation may be clockwise and is called as **right-handed rotation,** and the substance is called dextrorotatory or it may be counter clockwise called as **left-handed rotation** and the substance is called as levorotatory (Figure 7.2). Most of the biological molecules can rotate the plane polarized light so that they are optically active. For example, most of the sugars are dextrorotatory while the proteins and amino acids as well as phospholipids are levorotatory. The secondary and tertiary structures of proteins and DNA are asymmetric and so they are optically active.

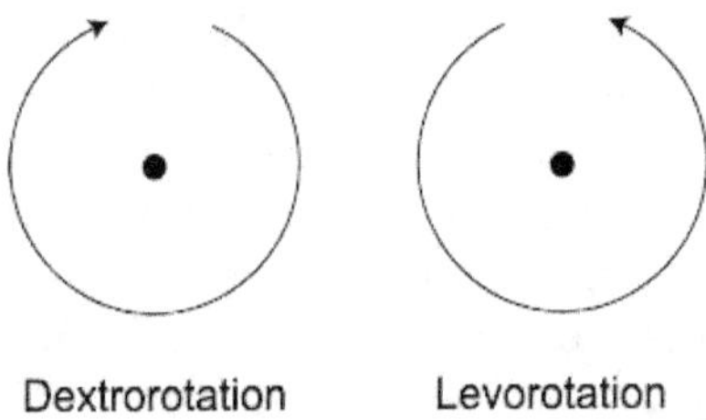

Figure 7.2. Rotation of light by optically active molecule.

7.1.2 Optical Rotatory Dispersion (ORD)

The optical activity of a substance is due to circular double refraction of light. In other words, the substance has a refractory index for right-circularly polarized light, which is different from the refractive index for left-polarized light. This means that the linear polarization of light is due to two circular polarizations of the same amplitude but with opposite rotation **(circularly birefringence).** The degree of rotation is proportional to the differences between two refractive indices and length of light velocities so that the emerging light beam will be out of phase with each other. Therefore, the light is rotated by an angle α to the original plane of polarization (Figure 7.3) and is dependent on wavelength. Thus at different wavelengths, the rotation of light is different and this effect is called **optical rotatory dispersion** (ORD).

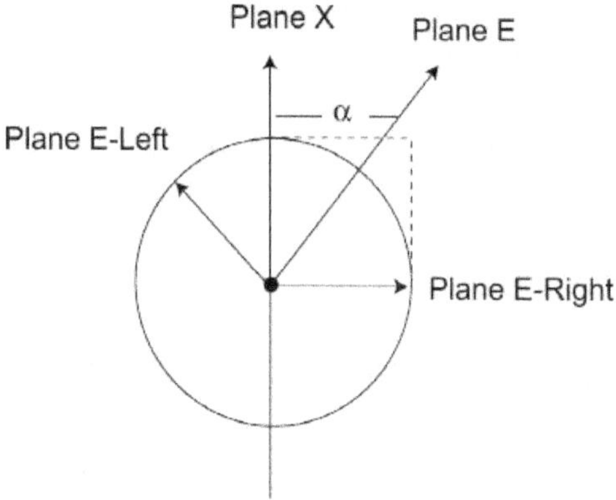

Figure 7.3. Rotation of polarized light from one plane to another.

7.2 CIRCULAR DICHROISM

The absorption of light by a molecule at a given frequency is different in different directions. Therefore the molecule appears to have different colours when viewed with two kinds of plane polarized light. This phenomenon is called **dichroism.** If the wavelength of the plane polarized light falls within the range of the absorption maxima of the medium, the rays absorbed by the medium will prefer either left or right circularly polarized rays. Therefore the transmitted rays have lesser intensity than the incident rays. This differential absorption of circularly polarized light is called **circular dichroism** (CD).

7.2.1 Measurements

ORD and CD are closely related phenomena so that CD spectrum of a sample can be derived from its ORD spectrum by mathematical calculations.

7.2.2 Measurement of ORD

A **spectropolarimeter** as shown diagrammatically in Figure 7.4 is used to measure the ORD spectrum. The apparatus has the following components.

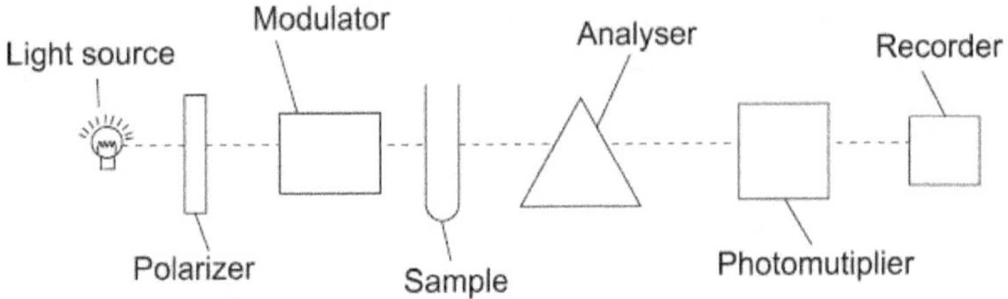

Figure 7.4. Diagrammatic structure of spectropolarimeter.

1. *Source of light* A tungsten filament lamp to get continuous light source or sodium or mercury vapour lamps to get light of specific wavelengths, are used as sources of illumination.

2. *Filter* The use of tungsten filament lamp requires sophisticated monochromators whereas sodium or mercury vapour lamp requires simple filters.

3. *Polarizer* It may be a Nicol prism, Glan-Thomson prism or a polaroid filter, which produces optical rotation.

4. *Analyser* It is a prism, which is used to measure the extent of optical rotation of the plane-polarized light emitted by the polarizer.

5. *Photomultiplier and recorder* The rotated plane polarized light is received by the photomultiplier and the recorder records the ORD spectrum.

7.2.3 Measurement of CD

A modified **spectrophotometer** with provision for the production of right and left circularly polarized light can be used to measure CD spectrum. The instrument is represented diagrammatically in Figure 7.5.

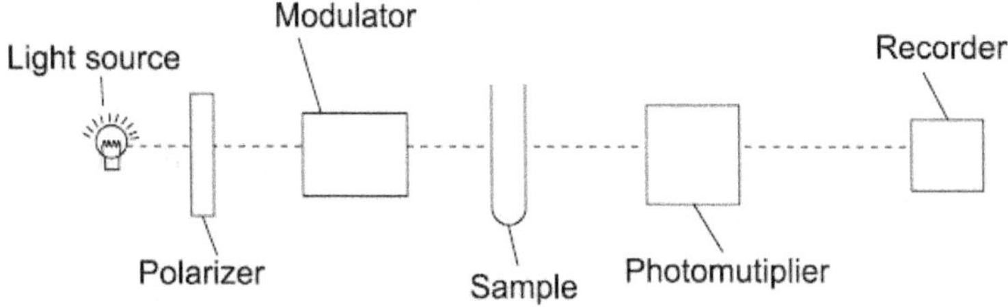

Figure 7.5. Diagrammatic structure of a CD recorder.

Here, the light rays are passed through a rotatable polarizer plate. The rotation of the plate through 45° at right or left side produces right or left circularly polarized light. The absorption of right and left circular rays through the sample is alternatively measured to get CD spectrum at a wavelength ranging from 185–600 nm.

Uses

- ORD and CD spectra are useful in the study of conformational changes in proteins and polypeptides. Proteins exhibit optical activity as they are formed by L-amino acids which are arranged orderly as α-helices and possess asymmetric distribution of charges in tertiary structure. Therefore, the general structure of protein molecules can be obtained through the determination of its optical activity.

- These methods are used to study enzyme–substrate complexes.

- In combination with ribose sugar, the bases in nucleic acids become optically active with highest activity in helical structures. Therefore, ORD/CD spectra give knowledge about structural conformation of nucleic acids, their denaturation and their binding with proteins.

- CD spectra are used to study the chain conformations of α-and β-linked glycan derivatives and polysaccharides. They are also employed to quantify optically active polysaccharides such as mucopolysaccharides.

7.3. LASER (LIGHT AMPLIFICATION BY STIMULATED EMISSION OF RADIATION)

If an atom is raised from its ground state (E_1) to an excited state or metastable level (E_2) through absorption of a photon, it is called induced absorption. An atom, which is initially in E_2 state, can drop to E_1 state by emitting a photon. It is called spontaneous emission. In addition, an excited atom can be made to emit a photon by another appropriate photon. Here, the atom in an excited energy state may come to lower energy state by the influence of the electromagnetic field of a photon and emit a photon of same frequency. In other words, the excited atom emits light in accordance with incoming waves resulting in increased light intensity (stimulated emission of radiation).

7.3.1 Principle

When the light of a particular frequency is passed through an assembly of atoms in the metastable state, then there will be more induced emission of light than induced absorption with the resultant amplification of light. Generally, the number of atoms is smaller at high-energy state and causes very little stimulated emission. When atoms are initially excited, there are more atoms at high-energy state. This phenomenon is called **population inversion**. This can be achieved by 'optical pumping' as shown in the Figure 7.6. Suppose if atoms of a material occur in three different states namely E_1, E_2 and E_3 and the atoms in ground state are pumped to E_3 state by photons, the excited atoms undergo non-radioactive transitions with the transfer of energy to E_2 state. Now there are more atoms at this metastable state and they produce stimulated emission when bombarded by photons.

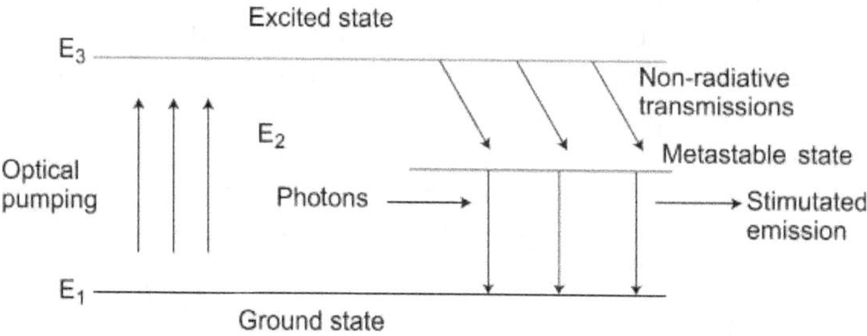

Figure 7.6. Diagram to show the phenomenon of optical pumping.

7.3.2 Instrumentation

The laser instruments are of two types namely **pulsed lasers (The Ruby laser)** and **continuous lasers (Helium–Neon laser)**. In pulsed lasers, the pumping radiation is stopped after getting the required number of molecules to the higher level and the laser produced is of a very short duration. In continuous lasers, the pumping radiation is uninterrupted so that there is a continuous renewal of molecules and the laser emission is continuous.

7.3.3 The Ruby Laser

The instrument has a cylindrical rod formed by aluminium oxide coated with chromium oxide with flat and parallel ends. One end of this tube is fully silvered and the end is partially silvered. These ends act as reflecting mirrors. The rod is externally surrounded by a glass tube, which in turn is surrounded by a helical xenon flash tube which acts as optical pumping system. The important component of the Ruby laser is shown in the Figure 7.7. From the xenon flash tube, the chromium ions absorb light and become excited from E_1 state to E_3 state. These ions now undergo non-radioactive transitions at E_2 level (metastable state) and produce spontaneous transition at E_1 level. Thus a beam of sufficient intensity is produced and emerges out as laser. However, laser production is stopped when the chromium ions return to the ground level. This necessitates another flash of pumping. But this problem is overcome by the use of helium–neon laser in which laser is produced continuously.

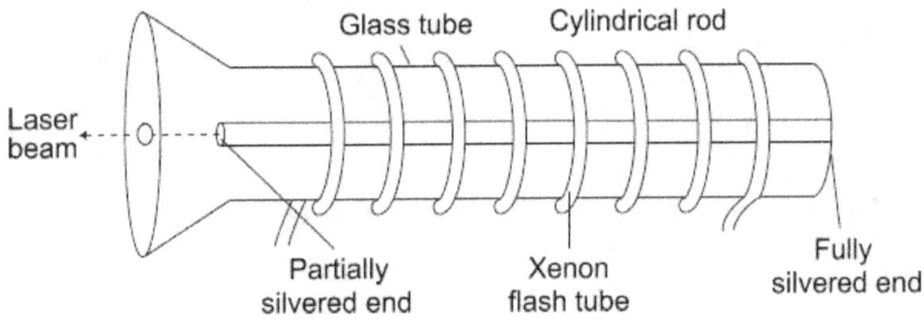

Figure 7.7 Components of a Ruby laser.

7.3.4 Helium–Neon Laser

As shown in the Figure 7.8 the helium–neon laser consists of a laser tube with parallel mirrors at the sides. One of the mirrors is fully transparent and the other one is partially transparent. The tube contains He–Ne mixture in a ratio of 5:1.

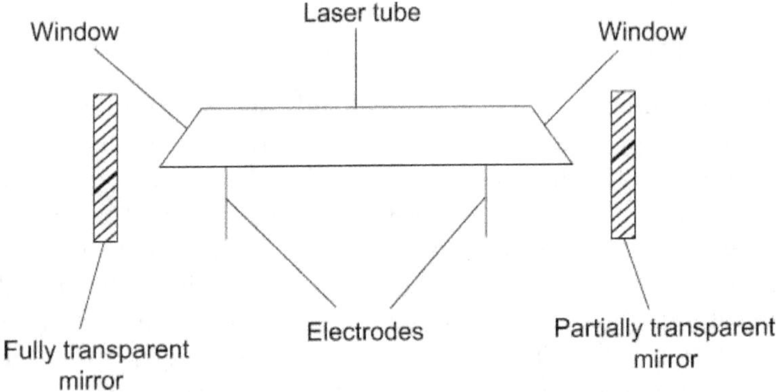

Figure 7.8. A Helium–neon laser (Diagrammatic).

The mixture is ionized by the electric current, which is applied through electrodes. On excitation, the He atoms cause population inversion in Ne atoms resulting in spontaneous transition which in turn triggers stimulated transitions. The mirrors help to reflect the photons and the windows prevent the reflection losses of photons. As the excitation of He and Ne atoms occurs uninterruptedly, He–Ne laser works in a continuous manner.

Uses   Laser technique is useful to produce three-dimensional optical images of molecules.

REVIEW QUESTIONS

1. What is optical rotatory dispersion (ORD)? How it can be measured?

2. What is circular dichroism (CD)? How can it be measured?

3. What is the principle of LASER?

4. What does optical pumping mean?

5. Describe the working principle of the Ruby laser.

6. Describe the working principle of Helium–neon laser.

8

BIOELECTRICITY AND NERVE IMPULSE CONDUCTION

The explanation of the action of living cells in terms of bioelectrical potentials constitutes **bioelectricity.** The body of higher organisms can be considered as a power station having two important sources namely muscles and nerves, which generate multiple electrical signals. These signals have a very low voltage and exhibit complicated pattern of electrical activity at cellular level. At resting state, the positively charged ions are distributed on the outer surface of the cell membrane and negatively charged ions inside. Thus unequal distribution of ions across the membrane causes a potential difference and thus a cell can be considered as a tiny 'biological battery'. In other words, the movement of positive and negative ions within a fluid medium produces the bioelectric current.

8.1 MEMBRANE POTENTIAL

The cytoplasm of living cells contains water, electrolytes and colloidal particles. Under an electric field, the electrolytes undergo ionic dissociation so that they become electrically charged. The colloidal particles are negatively charged and exhibit marked differences in ionic distribution due to Donnan equilibrium. The cell membrane is semi-permeable. Since the innermost ions of the cell are negatively charged and the ions outside the membrane are positively charged, the surface of the cell is positively charged and the inside of the cell negatively charged. At this stage, the membranc is said to be polarized and a potential difference occurs between the inner and outer surfaces of the membrane. This potential difference across the membrane is called **membrane potential**. Thus semipermeable nature of the membrane causes unequal concentration of both cations and anions in and out of the cell so that all living cells possess membrane potential.

At a resting state, potassium ions are more permeable than sodium ions whereas sodium ions are more permeable than potassium ions during excited state. Even though both anions and cations are found in equal concentrations inside and outside of the membrane, there is an excess number of anions on the inner surface of the membrane and cations on the outer surface due to active transport and diffusion of ions through the membrane. On stimulation, the permeability of the membrane is increased at the point of stimulation where the membrane becomes negative to the adjacent point (**depolarization**). The removal of stimulus leads to ionic equilibrium and the membrane becomes repolarized. These phenomena of depolarization and repolarization are collectively called as **membrane phenomenon**. In this way, the unequal distribution of charged particles produces differences in electrical potential, which is responsible for the functioning of living organisms.

8.2 RESTING MEMBRANE POTENTIAL

In a resting state, there is no difference in electrical potential on either side of the membrane and the membrane is said to be in an isopotential state. Though the resting membrane is in an isopotential state, the proximity of external surface membrane contains a slight excess of cations over anions whereas the proximity of inner surface membrane, a slight excess of anions over cations due to ionic influx. Because of these slight changes in ionic distribution, a voltage difference of –75 to 95 mV occurs at the two surfaces of the membrane and is called **resting membrane potential**. The resting potential is maintained by the active transport of ions against electrochemical gradient by sodium/potassium pumps. During resting state, the concentration of potassium ions is higher (140 mE/l) and that of sodium ions are lower (10 mE/l) inside the membrane. The fluid outside the membrane has higher concentration of sodium ions (142 mE/l) and lower concentration of potassium ions (5 mE/l). These differential ionic distributions are due to sodium and potassium pumps by diffusion and active transport. Therefore the outside membrane becomes positively charged and the inside membrane becomes negatively charged because of the presence of higher concentrations of negatively charged ions such as chlorides and proteins. In a resting state of the nerve, there is an active transport of sodium ions from the axoplasm into the interstitial fluid. This transport of sodium ions is known as **sodium pump.** As a result, the interior of the axon becomes strongly electronegative. This is the normal polarized state of the membrane of the resting nerve fibre. (Figure 8.1).

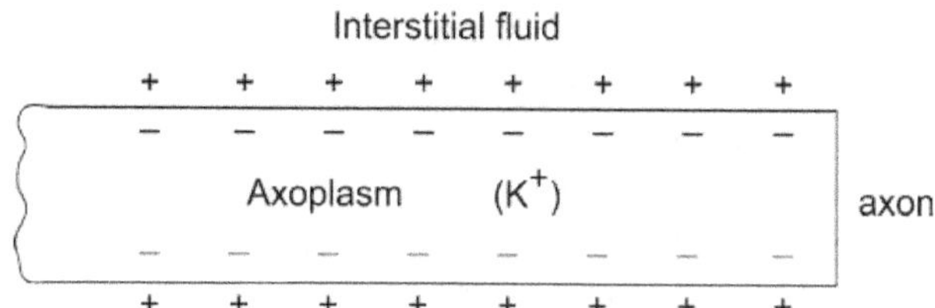

Figure 8.1. Polarized state of the membrane in a resting nerve fibre.

8.3 ACTION POTENTIAL AND NERVE IMPULSE CONDUCTION

In most animals, coordination is effected by means of neurons or nerve cells, which are capable of transmitting or conducting the impulses in the form of electrical wave through specific pathways. According to Prosser, the sum of total physical and chemical reactions in the conduction of a wave of physiological activity along the nerve fibre is called an **impulse**. When the nerve is subjected to external stimuli, it is excited and sets up a wave of impulse, which travels along the nerve fibres.

The action potential occurs in a nerve when it is stimulated by a stimulus, which may be mechanical, electrical or chemical. On applying a stimulus to a **non-myelinated nerve,** the membrane of the nerve fibre becomes depolarized and the sodium pump stops. The permeability of the membrane for the sodium ions increases so that the sodium ions from outside the membrane tend to move into the axoplasm. This causes a change in the polarization of the membrane and the interior of the fibre becomes positive while the outside becomes negative. This event is just opposite of the resting potential and is called **depolarization** or **reversal potential** (Figure 8.2). The prevalence of electropositivity inside and electronegativity outside the excited nerve fibre is called **action potential**. The flow of sodium ions from the original depolarized area proceeds further in both directions. Thus more and more sodium ions enter the axoplasm until the whole interior of the nerve fibre becomes electronegative. This movement of sodium ions as an electric current, which spreads along the membrane, is called **depolarization wave** or **nerve impulse**. As soon as the depolarization wave spreads completely over the entire length of the fibre, sodium ions cannot enter further into the axoplasm. But a large quantity of potassium ions diffuses through the membrane into the axoplasm. Now the sodium ions begin to diffuse out simultaneously with the resultant electronegativity inside the membrane and electropositivity outside. This process occurs at the same point of the nerve fibre in which depolarization has started and extends in both directions (Figure 8.3). This is called repolarization during which the sodium pump starts once again.

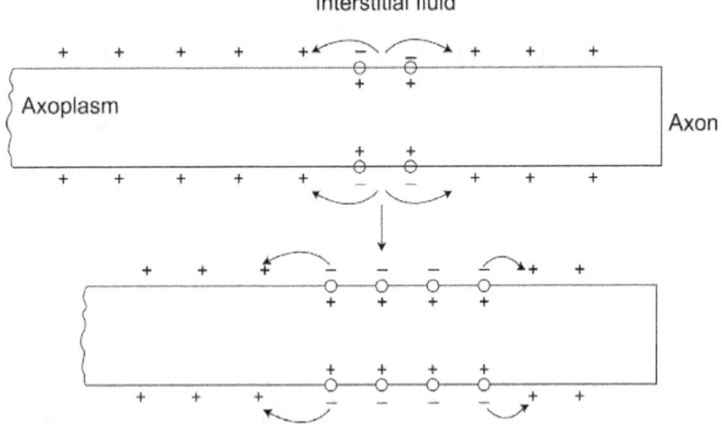

Figure 8.2. The event of depolarization in a stimulated nerve fibre.

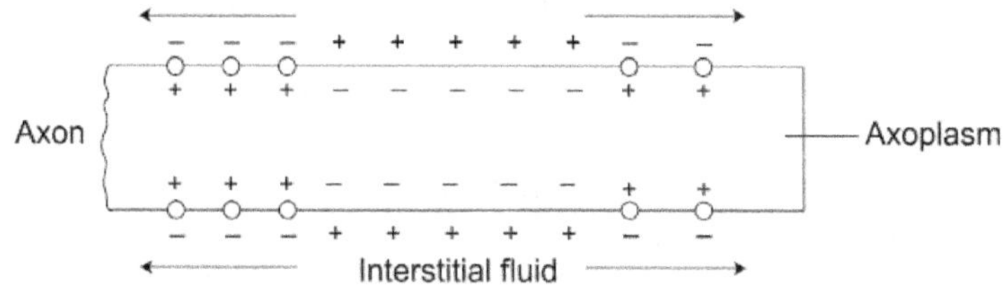

Figure 8.3. Repolarization in a nerve fibre.

The whole process of depolarization and repolarization is completed within a small fraction of a second and thus the nerve fibre can conduct impulses very rapidly.

In vertebrates, the efficiency of impulse conduction is increased by **myelin sheath**, which surrounds the axon, acting as an effective insulator. In **myelinated nerves** also, the action potential arises in the same manner as in the non-myelinated nerves but only at the **nodes of Ranvier**, which lack the sheath. Here, the action potential actually jumps from one node to another and the process is called **saltatory conduction**. In myelinated nerves, the current spreads outwards from the node so that the action potential reaches the next mode from its origin faster than the impulse, which would travel along the same distance in a non-myelinated nerve fibre. Therefore the velocity of conduction depends on the diameter of the nerve fibres.

In **synapses,** the impulse transmission is mediated by a neurotransmitter substance (**acetyl choline**). When a nerve impulse reaches the synaptic knob, it depolarizes the presynaptic membrane and causes the permeability of the membrane to calcium ions. Now the calcium ions enter into the knob and cause the fusion of synaptic vesicles with the presynaptic membrane. The vesicles release their contents (neurotransmitter substance) into the synaptic cleft by exocytosis. Then they return to the cytoplasm for refilling. The released acetylcholine diffuses across the synaptic cleft, attaches to a specific receptor site on the postsynaptic membrane and causes a change in the shape of the receptor site. This in turn results in opening of ion channels in the postsynaptic membrane. Thus ion channels are opened in response to binding of acetylcholine to receptor proteins. As a result, sodium ions which enter through the postsynaptic membrane causes depolarization of the membrane. This excites the cell to develop action potential (nerve impulse). The acetylcholine is immediately removed from the synaptic cleft by the enzyme acetyl cholinesterase present on the postsynaptic membrane. The synaptic transmission of the impulse is schematically shown in the Figure 8.4.

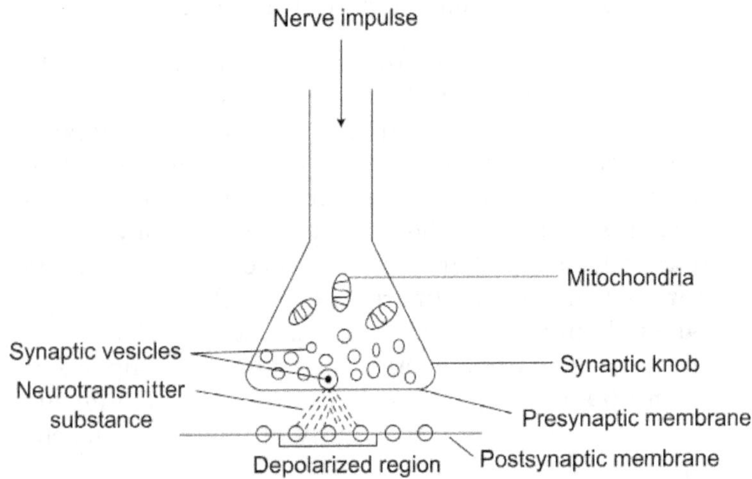

Figure 8.4. Synaptic transmission of a nerve impulse (Diagrammatic).

The **neuromuscular junction** is a specialized synapse found between a motor neuron and skeletal muscle fibres. It includes both motor end plate and the synaptic knob of the neuron (Figure 8.5). The fine branches of neuron terminate in shallow cavities of sarcolemma (cell surface membrane of the muscle fibre), which has many deep folds containing receptors for acetylcholine in the cavities. On stimulation, the synaptic knobs release acetylcholine by the same mechanism as in synapses (Figure 8.6). Change in structure of receptor sites on the folds of sarcolemma (postsynaptic membrane) increases the permeability of the sarcolemma to sodium and potassium ions resulting in a local depolarization called **end-plate potential**. This potential causes an action potential, which is transmitted along the sarcolemma to the muscle fibre through the T system of the muscle fibre.

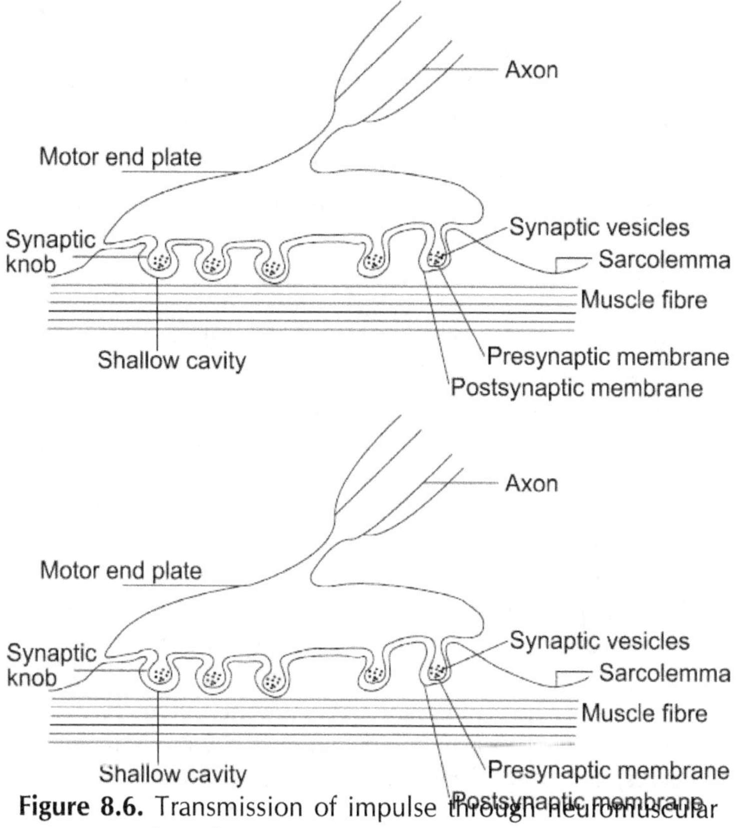

Figure 8.6. Transmission of impulse through neuromuscular junction.

8.4 RATE OF NERVE IMPULSE CONDUCTION

A nerve impulse is an action potential passing along the axon as a wave of depolarization. At the site of action potential, the outside surface of the axon is negatively charged. Action potential is generated by the entry of sodium ions into the axon thus creating an area of positive charge. Therefore, a flow of current is set up in a local circuit between this active area and the negatively charged resting region immediately ahead. This flow of current reduces the membrane potential in the resting region leading to depolarization. In turn, the depolarization produces an increase in sodium permeability with the resultant action potential in this region. The repeated depolarization of immediately adjacent regions of the membrane results in the propagation of action potential along the axon.

Action potentials are capable of being transmitted over an infinite distance. The size of the action potential is not affected by the intensity of the stimulus (all or none law). It means that there is a threshold stimulus intensity above which the action potential will be triggered whatever may be the strength of the stimulus. The speed at which the impulse is transmitted is also not affected by the intensity of the stimulus. However, the frequency of action potentials is directly proportional to the strength of the stimuli within limits. Most nerve fibres can conduct at a rate of 10–100 impulses per second. Some can also conduct as many as 500 impulses per second. In non-myelinated axons, the speed of conduction of action potential depends on the resistance of the axoplasm which in turn depends on the diameter of the axon. For example, the axons with smaller diameter (< 0.1 mm) possess greater resistance, which reduces the length of the local circuits. The speed of conduction in these fibres is 0.5 pulses per second. On the other hand, in giant axons (1 mm diameter), the velocity of conduction is up to 100 impulses per second. In myelinated nerves, the speed of conduction is up to 120 impulses per second.

8.5 RECORDING OF NERVE IMPULSES

The bioelectric current produced by the cells can be recorded with suitable devices. On excitation, the charges tend to migrate through fluids towards unexcited cell areas causing electric current. This produces potential differences in various regions of the cell. These differences are measured by placing electrodes at two different points and recorded with the help of electric instruments. The electrodes play an important role in recording the bioelectric signals and they are of two types namely **surface electrodes** and **deep-seated electrodes.** The surface electrodes measure the potential difference on the surface of the tissue without damaging the cell. The deep-seated electrodes measure the electric potential, which occurs in the interior of the cell. **Microelectrodes** made of metals or glasses are used to study the electrical activity in individual cells. The signal from the microelectrodes are sent to an ordinary galvanometer or to a sophisticated instrument called **cathode ray oscilloscope.**

Structure of cathode ray oscilloscope   Various parts of a cathode ray oscilloscope are shown in the Figure 8.7. It has a cathode ray tube consisting of an electron gun formed by two electrodes (a thread-like cathode and a disc-like anode with a small aperture in the middle) placed in a vacuum glass tube and two sets of plates with a fluorescent screen at the base of the tube. Of the plates, two are horizontal deflection plates placed on either side of the path of the electron beam and the other two are vertical deflection plates placed above and below the beam.

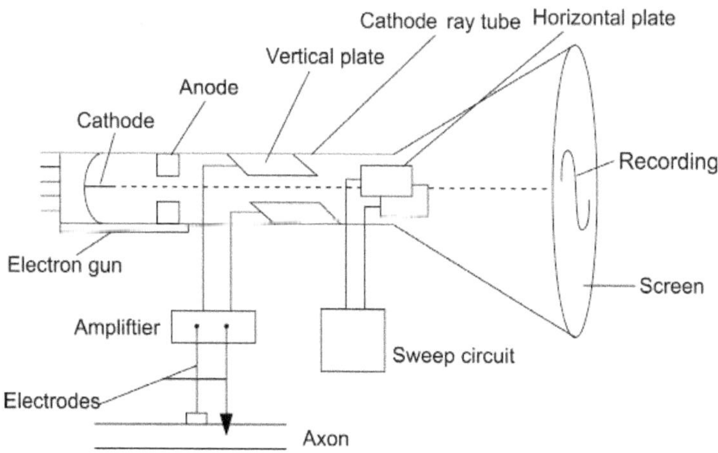

Figure 8.7. Structure of a cathode ray oscilloscope.

The potential difference across the vertical plates may cause a deflection of negatively charged electrons and the horizontal plates make the electron beams to move horizontally across the tube. The electrical circuit includes a sweep circuit to control the current intensity on the horizontal plates and an amplifier to control the intensity of the current on the vertical plates and to amplify the electrical wave. When high voltage is applied to the electrodes, the cathode produces a stream of free electrons, which pass through the aperture of the anodal disc, and falls on the screen producing a luminescent spot. If the electron beam is moved across the surface of the screen, the spot also moves and produces a fluorescent line on the screen. When a negative charge is applied to the left vertical plate and a positive charge to the right plate, the electron beam repels away from the left plate and attaches to the right plate. As a result, the spot of light moves to the right on the screen. Thus positive and negative charges are used on the vertical plates to move the beam up and down.

8.5.1 Resting Membrane Potential

In a resting state, if two electrodes are placed on the surface of the membrane of a nerve, there is no current flow. This shows the absence of differential electric potential. There is no deflection in a galvanometer and the membrane is said to be in an isopotential state (polarized membrane) as shown Figure 8.8.

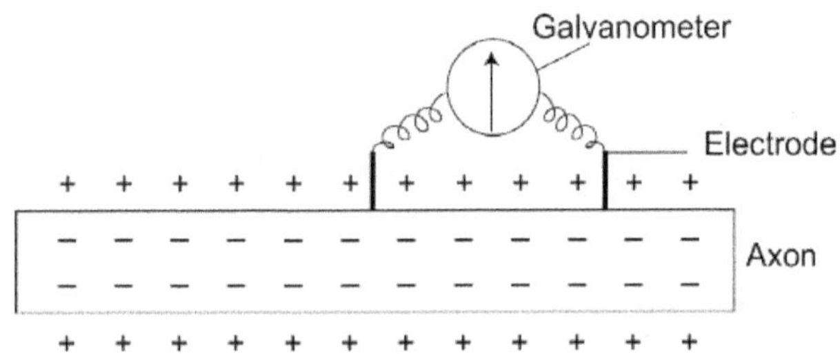

Figure 8.8. Isopotential state of a resting membrane.

8.5.2 Injury Potential

When one of the electrodes is applied to the surface of the resting part of the nerve fibre and another injured part as shown in Figure 8.9, there will be flow of current due to the positivity of the uninjured site and negativity of the injured site and the galvanometer deflects to right. This potential is called **injury potential** or **demarcation potential.**

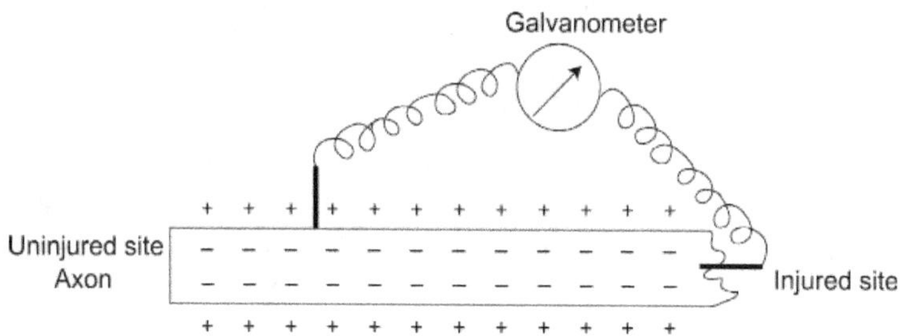

Figure 8.9. Injury potential in a nerve fibre.

8.5.3 Monophasic Action Potential

To record the monophasic action potential, a nerve fibre is injured at a point x which becomes negative as in the Figure 8.10. On applying a single stimulus to a point z, when one electrode is placed at the point y to which the contraction wave reaches and the other electrode on the injured site, then the galvanometer shows no deflection. This is because the site at which the stimulus is given as well as the injured site exhibit electronegativity so that there will be no potential difference. When the contraction wave moves away from point y, repolarization occurs in the membrane and point becomes electropositive. This results in differential electrical potential and the galvanometer exhibits only a single deflection. This is called **monophasic potential.**

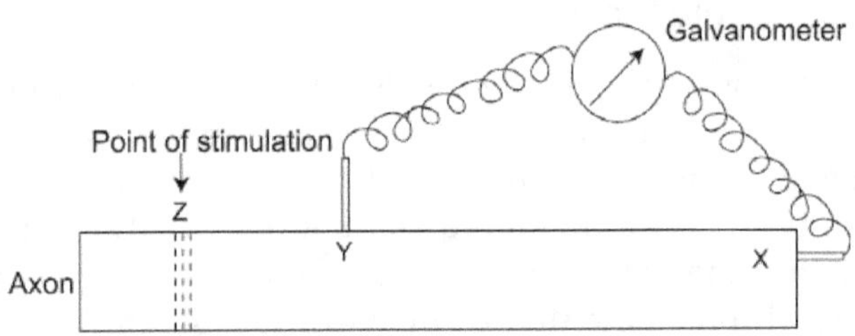

Figure 8.10. Monophasic action potential in a nerve fibre.

8.5.4 Diphasic Action Potential

To record diphasic or biphasic action potential, the surface of an uninjured nerve fibre is connected to a galvanometer by two electrodes. When a stimulus is applied to the fibre, the contraction wave exhibits two deflections when it is transmitted along the nerve. This is called diphasic action potential, the events of which are shown in Figure 8.11.

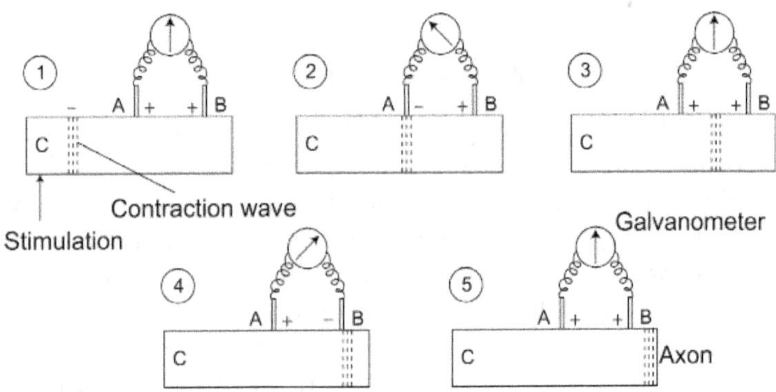

Figure 8.11. Events during diphasic action potential.

On stimulation, the point C is depolarized and become negative whereas the points A and B are positive. Therefore the meter shows no deflection (Figure 8.11) when the wave of contraction reaches the point A, it becomes negative and the point B is positive, so that the meter shows deflection to the left (Figure 8.11). When the wave comes to the point between A and B, this point becomes negative whereas the points A and B are positive, the result being no deflection (Figure 8.11). When the wave reaches the point B, it becomes negative and the point A is positive so that the meter deflects to the right (Figure 8.11). When the wave moves away from the point B, then this point becomes positive and the point A is also positive. Therefore there is no deflection in the meter (Figure 8.11).

The recording of a diphasic potential by cathode ray oscilloscope is shown in Figure 8.12. It shows that there is an upward deflection followed by an isoelectric interval and then a downward deflection. This indicates that the potential difference is in one direction at first and then in opposite direction. This sequence of events is called **diphasic action potential.**

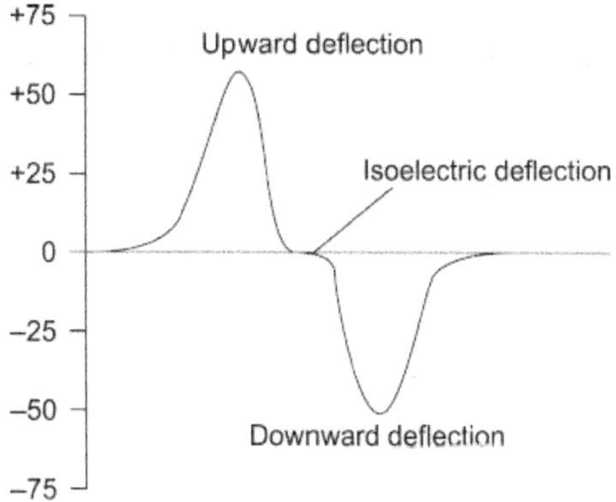

Figure 8.12. Record of a diphasic action potential by cathode ray oscilloscope.

REVIEW QUESTIONS

1. Define bioelectricity, resting membrane potential, action potential and nerve impulse.

2. Differentiate between depolarization and repolarization.

3. Explain the membrane phenomenon.

4. What is the nature of resting membrane potential?

5. What is the nature of action potential?

6. Describe the process of nerve impulse conduction in a non-myelinated nerve fibre.

7. Explain synaptic conduction of nerve impulses.

8. Explain the mechanism of impulse conduction at neuromuscular junction.

9. Write a note on the rate of nerve impulse conduction.

10. What are different types of electrodes used to record nerve impulses?

11. Describe the structure of cathode ray oscilloscope.

12. How could resting membrane potential be measured?

13. How could injury potential be measured?

14. How could monophasic action potential be measured?

15. How could diphasic action potential be measured?

RADIATION BIOLOGY

9.1 RADIOACTIVITY

An atom consists of a positively charged nucleus with positively charged protons and neutral neutrons surrounded by electrons. In an atom, the number of electrons is always equal to the number of protons making the atoms electrically neutral. The number of electrons in an atom denotes its atomic number and the number of protons and neutrons together represents the mass number. Atoms, which have the same atomic number but different mass numbers are called **isotopes.** A stable isotope is one in which the nucleus has equal number of protons and neutrons. If the proton-neutron ratio of atoms is more than one, they become unstable and are called **radioisotopes**. The process of emission of electromagnetic radiation by the radioactive isotopes is called **radioactivity**, which may be natural or artificial (man-made).

9.1.1 Natural Radioactivity

The spontaneous emission of highly penetrating radiations such as α, β and γ rays by the heavy elements like uranium, radium, thorium, etc. is called **natural radioactivity**. These elements are called **radioactive elements**. The α particles are relatively larger in size, positively charged and are emitted by the nucleus. They contain two protons and two neutrons. They pass through straight lines with high velocity. They can ionize gases, affect photographic plates and produce fluorescence. The β rays are similar to electrons as they carry a negative charge. They are smaller than α particles but their velocity is greater approaching that of light. β rays exhibit two types of spectra namely line spectra and continuous spectra. The line spectra are produced by the emission of electrons from the orbits of atoms whereas the continuous spectra are produced by the emission of electrons from the nucleus of atoms. γ rays are similar to X-rays and are nuclear in origin. They have very short wavelengths and are capable of emitting a large quantum of energy. The process of elimination of α, β and γ rays is called **particulate** or

corpuscular radiation whereas the elimination of small units of energy namely photons or quanta are called **electromagnetic radiation**. Particulate particles travel at a speed less than that of light. On the other hand photons travel at the speed of light.

9.1.2 Artificial or Induced Radioactivity

The process by which the light elements are made into radioactive substances is called **artificial radioactivity.** These artificially induced substances also exhibit decay similar to natural radioactive elements but they emit electrons, neutrons, positrons and α rays. Almost all elements can be made into radioactive substances by bombarding them with particles like α particles. The important artificial radioisotopes include tritium (H_3), sodium (Na^{24}), phosphorous (P^{32}), cobalt (Co^{60}), carbon (C^{14}), iodine (I^{131}), sulphur (S^{35}), etc.

9.1.3 Radioactive Disintegration

When an unstable atom emits radiation, it changes either in structure or energy level or both and this change is called **radioactive disintegration** or **radioactive decay**. The process by which the radioactive substances disintegrate to form a series of new substances is called **successive disintegration**. After the decay, some parent substances exist so that the lifetime of radioisotopes is referred to as **half life**, which is the time taken for the disintegration of one half of the radioactive substances (rate of disintegration).

9.1.4 Units of Radioactivity

In general, the radioactivity of a substance is expressed in terms of Curie (Ci). A curie is the quantity of a radioactive substance, which gives 3.70×10^{10} disintegrations/sec. Another unit is Roentgen (r), which is the quantity of radiation producing ions equivalent to one electrostatic unit of electricity through a volume of air. The effect of radioactivity in biological systems is expressed as Rad (Roentgen absorbed dose) and rem (Roentgen equivalent man). One rad is equal to 100 ergs per gram and one rem is approximately equal to 1.00 rad for β, γ and X-rays and to 0.05 rad for α rays.

The concentration of naturally occurring isotopes is expressed as atoms percent (%) excess. The concentration of isotopes, which are not occurring naturally, is expressed as the specific activity, i.e. number of atoms/unit of the substance (mg, μ mole etc.). The distribution of isotopes in the precursor and product can be determined by 'dilution factor', which is the ratio of specific activity of the precursor to specific activity of the product. In experiments which use radioactive isotopes, the incorporation of radioactivity is denoted as counts/minute/gram.

MEASUREMENT OF RADIOACTIVITY

The radioactivity of isotopes can be measured either by using counting devices or by photographic method (autoradiography).

The quantitative determination of radioactivity is based on the ionization or excitation of matter induced by radiations emanated from radioactive elements. The following are the important instruments used to determine radioactivity of substances.

9.2 GEIGER-MULLER COUNTER (GM TUBE)

This instrument works on the principle of ionization of gases and is suitable for all kinds of radiations. The structure of GM tube is schematically shown in the Figure 9.1. It consists of a simple and very compact ionization chamber made of glass with two electrodes. It is filled with a gas usually argon, neon or air. The outer electrode (cathode) is a metal tube formed by brass or nickel with 1–5 cm diameter and 10–50 cm length. The inner electrode (anode) is a fine wire of tungsten with

0.1–0.5 mm thickness. The anode is stretched along the axis of the tube and well-insulated by ebonite plugs. When radiation enters the tube, the gas is ionized rapidly under a high voltage power supply resulting in an electric pulse. But, the discharge of the pulse terminates when large quantity of positive ions get collected around the negative electrode and this prevents the occurrence of further ionization. To accelerate the process of ionization, GM tubes contain a quenching gas. However, presence of a quenching gas and the window through which the particles pass into the tube restrict the entrance of radiation thereby reducing the efficiency of the detector. To overcome this problem, GM tubes with very thin windows or with no windows have been developed. But the windowless GM tubes require a continuous flow of special quenching gas through the tube.

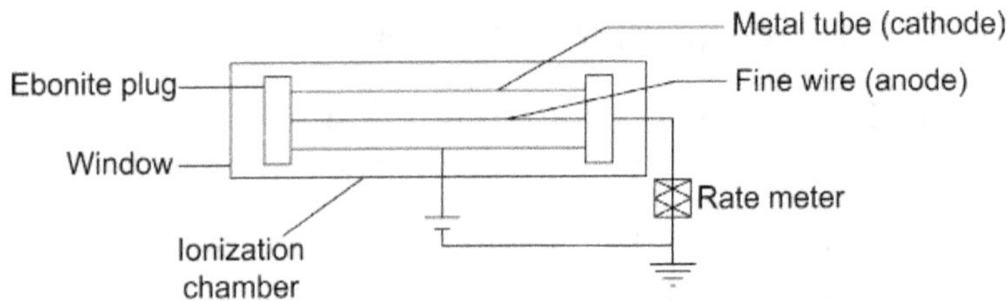

Figure 9.1. Diagrammatic structure of GM tube.

9.3 PROPORTIONAL COUNTER

It is a modified GM tube which operates in a region of applied current where the charge produced is proportional to the initial ionization. This device is more advantageous than GM counter because it has greater stability and reproducibility and can operate at reduced voltage. The consumption of quenching gas is lesser in proportional counter than in GM tube. This device produces continuous pulses which enable very fast counting. The most popular proportional counter is the 'flow counter' in which a gas like argon or indane is continuously passed through an open-ended tube of ionization chamber at a pressure slightly higher than that of the atmosphere.

9.4 SCINTILLATION COUNTER

This apparatus is more sensitive and efficient than the GM tube and is mostly used in detecting γ-rays. It uses certain fluorescent crystals, which emit light in the form of scintillations on absorbing radiation. The main parts of this counter are schematically shown in Figure 9.2. Phosphor is usually used as scintillating substance. Natural inorganic crystals (Zinc sulphide or calcium tungstate) and certain organic liquids as well as plastics are also used as phosphors. Inorganic phosphors are mainly used to detect protons, deuterons and γ rays whereas organic phosphors are used to detect α particles. These substances when exposed to radiation scintillate and emit flashes of light. These light pulses are converted into electric pulses, which are amplified and recorded.

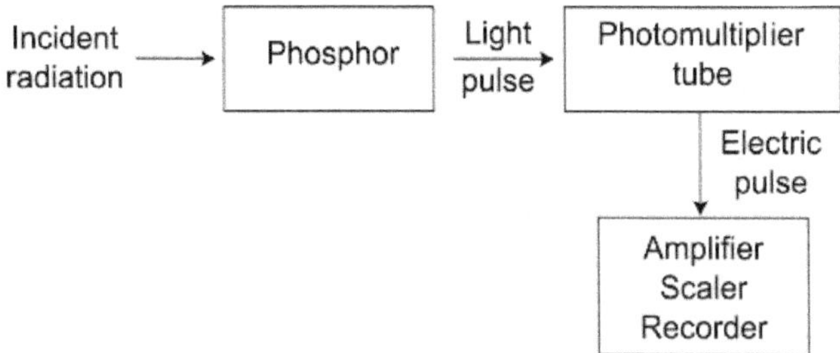

Figure 9.2. Schematic representation of scintillation counter.

9.5 LIQUID SCINTILLATION COUNTER

This device is highly efficient in counting very weak β-rays. Here the radioactive material is dissolved in a liquid containing a phosphor which interacts with β particles resulting in the emission of light flashes.

Recently many types of scintillation counters have been devised to study cosmic rays.

9.6 CRYSTAL COUNTER

In this radioactive counter, photoconductivity of crystals is utilized to determine the radioactivity of substances. The diagrammatic structure of a crystal counter is shown in figure 9.3. It consists of a natural crystal (diamond or silver chloride or lithium fluoride), which is mounted between two metal electrodes maintained at high voltage. Nowadays semiconductors like silicon crystals are also used. When radiation is allowed to pass across the crystal, electrons are expelled from the atoms of the crystal and accelerated towards the positive electrode. On their way, the accelerated electrons hit more electrons. As a result, a bunch of electrons reach the electrode with the production of a corresponding pulse.

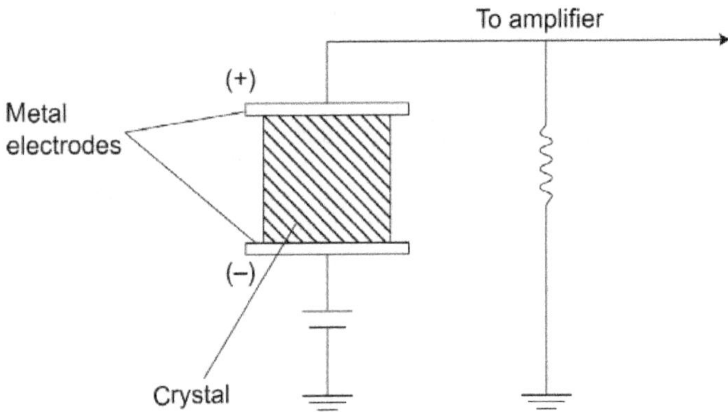

Figure 9.3. Diagrammatic structure of crystal counter.

9.6.1 Method of Detection of Disintegration Frequency

In all detectors, the disintegration frequency of radioactive substances is counted either by a scaler or by a count rate meter. A scaler is an electronic device, which counts the number of pulses from the counter equipments. The count rate meter converts the counts into counts per minute, which are read directly. A spectrophotometer is also used between the counter and scaler or rate meter to differentiate pulses of different amplitudes.

9.7 AUTORADIOGRAPHY

Autoradiography is an important technique by which the routes and conversion of molecules in biochemical reactions occurring in cells are traced. The radioactive isotopes are stable only for definite periods and then they decompose to form other atoms or fragments of atoms. During this radioactive decay, many atoms produce radioactive emissions, which may be measured by special devices.

In this technique, the sample containing radioactive substance (a chromatogram or a tissue slice or a section of a plant or animal) is kept in close contact with a photographic plate or an X-ray film. That is, the cells or tissues already exposed to radioactive isotopes are made to contact with a photographic emulsion for specific period as shown in the figure 9.4. The emanating radiations affect the emulsion in the same way like that of light. On development, the areas, which have been in contact with radioactivity, could appear as spots on the film, thus producing the image of the specimen.

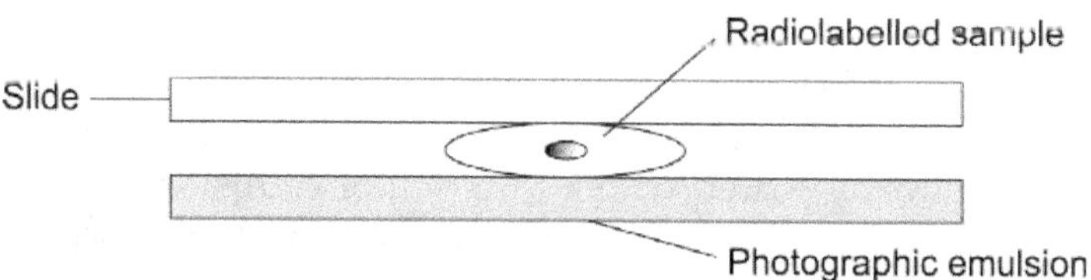

Figure 9.4. Contact between radiolabelled sample and photographic emulsion.

9.7.1 Specimen Preparation for Autoradiography

Typically, small precursor molecules, which are utilized by the cells as building blocks of macromolecules, are incorporated with radioactive isotopes. This is usually carried out by injecting suitable amount of a dilute radioactive solution into the animal (if required, repeatedly). Then the animal is anaesthetized and a specific tissue is dissected out after a period of time for fixation. The choice of fixation and fixatives depends on the chemical composition of the sample material to be studied. For example, nucleoproteins are fixed in Carroys acetic alcohol. The tissues are quickly frozen and dried in vacuum desiccators to obtain all the incorporated radioactive molecules. The incorporation of a radioactive isotope into RNA of cells requires the extraction of all DNA molecules from the tissue. Similarly, the study of DNA necessitates the removal of all RNA molecules from the tissues.

For tissues, the usual paraffin sections of 2 μ thickness are made by routine microtechnique method. Smears and frozen sections can also be radiolabelled and studied.

9.7.2 Methods to Make Contact Between Specimen and Photographic Emulsion

In general, three methods are employed to make contact between the specimen and photographic plate.

i) *Opposition method* This method is used to study single cells and the contact of the specimen with photographic emulsion is made in dark. First, the paraffin sections are allowed to float on the warm water and then picked up on a microscopic slide, which is already coated with a thin film of emulsion. Now the sections remain attached to the slide and are processed, stained and mounted.

ii) *Liquid–emulsion method* Here, the slides with deparaffinized sections, cell smears or blood films are simply dipped in molten liquid emulsion, dried in a light-tight box and kept in refrigerator for desired period of exposure. Then the slides are developed in dark room, fixed, stained and examined under phase-contrast microscope.

iii) *Stripping–film method* The special stripping film meant for autoradiography is cut into a square in dark room and is floated in distilled water contained in a cylindrical glass jar with the emulsion side facing downwards. The deparaffinized slide containing the sections is immersed in water below the filmstrip and lifted out of water in such a way that the film covers the specimen. Then the slide is dried, stored in dark for exposure, developed, fixed, stained and observed under phase contrast microscope.

Uses Autoradiography is useful in the study of the movement of solutes in tissues, to determine the rate of production and life span of blood cells and to identify thyroid

Table 9.1. Contributions of autoradiography to basic cell metabolism.

Radiolabelled substances	Events in cell metabolism
C^{14} labelled glycine	Synthesis of RNA from nucleolus, protein synthesis
H^3 labelled thymidine	Metabolism of chromosomal DNA, cell turnover times, DNA metabolism in salivary gland chromosomes
H^3 labelled cytidine	Nucleic acid metabolism during regeneration
H^3 labelled cytidine and thymidine and C^{14} labelled glycine	Changes in nucleic acids and protein metabolism during meiosis in plants
C^{14} labelled adenine	nucleic acid metabolism during animal spermatogenesis

function and cancer cells. This technique has made many contributions in the study of cell metabolism as given in Table 9.1. In medicine, this technique is useful to diagnose circulatory disorders, bone defects, tumours and so on. Some like I^{131} is used to treat thyroid defects, P^{32} to treat skin diseases and Co^{60} to treat tumours. In agriculture, the incorporation of radiolabelled fertilizers to the soil increases crop yields. Radioisotopes are also used to preserve vegetables and cereals. Radiation is used to sterilize pharmaceutical and surgical instruments. The isotope of carbon is useful in archaeological dating to estimate the age of earth.

9.8 BIOLOGICAL EFFECTS OF RADIATION

α, β and γ rays as well as neutrons cause harmful effects on living organisms due to ionization of atoms with in the cells. In other words, ionization alters or destroys cellular constituents and causes nuclear-cytoplasmic swelling, breakage of chromosomes and so on. In general, the effects of radiation can be categorized into three types. The **recoverable effects** or **short-term effects** are due to small doses of radiations and include skin disorders and loss of hair. (**Short-term recoverable effects**). The **irrecoverable effects** or **long-term effects** are due to larger doses of radiations and include bone marrow damage, cell damage, leukemia, cancer, etc. (**long-term irrecoverable effects**). The **genetic effects** appear only in future generations of irradiated individuals and cause deterioration of quality at population level by way of producing unusual offsprings.

REVIEW QUESTIONS

1.　What is radioactivity? Classify radioactivity with examples.

2.　What is radioactive disintegration?

3.　Write a note on units of radioactivity.

4. Describe the structure and functioning of Geiger-Muller counter.

5. Describe the structure and functioning of proportional counter.

6. Describe the structure and functioning of scintillation counter.

7. Describe the structure and functioning of liquid scintillation counter.

8. Describe the structure and functioning of the crystal counter.

9. What is autoradiography? Mention its principle.

10. Enumerate the steps involved in preparing the specimen for autoradiography.

11. In what ways is the contact between the specimen and photographic plate made in autoradiography?

12. Write an essay on major contributions of autography in biological studies.

13. Explain various biological effects of radiation.

GLOSSARY

Action potential   the prevalence of electropositivity inside and electronegativity outside the excited nerve fibre.

Adsorption   the process of adherence of foreign molecules to a solid or liquid surface.

Adsorption spectrum   the spectrum obtained by the interaction in which the substance absorbs radiation.

Amphipathic groups   possess both hydrophilic polar groups and hydrophobic nonpolar groups.

Anion   negatively charged ion that will move towards the anode of an electrolytic cell.

Bioelectricity   the explanation of the action of living cells in terms of bioelectrical potentials.

Bond distance   the equilibrium distance between the nuclei of two atoms.

Bond energy   the energy released at the time of overlapping of atomic orbitals to form bonds.

Buffer solution   a solution that will resist to changes in pH on addition of small quantity of acid or alkali.

Cation   positively charged ion that will move towards the cathode of an electrolytic cell.

Centrifugal force   the radial outward force acting on a body that is moving in a circle with a particular radius and speed.

Conformation   refers to the arrangement of atoms due to rotation and orientation of part of the molecule.

Covalent bond   a strong interaction in which two atoms share one or multiple pairs of electrons.

Crystals　three-dimensional space gratings used for the diffraction of X-rays.

Dialysis　the process of retention of colloidal particles by a semipermeable membrane and the movement of dispersion medium.

Dichroism　the appearance of a molecule to have different colours when viewed with two kinds of plane-polarized light.

Diffusion　the movement of ions and molecules across a membrane from their higher concentration to lower concentration without requiring energy.

Dipole　a system consisting of two opposite electric charges separated by a distance.

Dipole moment　the potential developed by the electric symmetry in a dipole.

Electric polarization　the displacement of negative and positive charges in an atom under the electric field.

Electrode　a device which makes surface contact between metallic and non-metallic conductors.

Electromagnetic radiation　a form of energy transmitted through space with enormous velocity without requiring any supporting media.

Electron gun　a heated cathode that produces a stream of electrons in vacuum.

Electron spin resonance　the absorption of radiation by the electrons in a magnetic field and their change from low-energy state to high-energy state.

Emission spectrum　the spectrum obtained by the interaction in which a substance itself emits radiation.

Filter a device used to allow only certain frequencies to pass through and to prevent other frequencies.

Filtration the passage of fluid through a membrane due to differential hydrostatic pressure.

Fluorescence the absorption of light by a molecule with a specific wavelength and its emission at a different wavelength.

Focal length the distance between the focal point and the optical system.

Hydrophilic groups groups that have orientation towards water molecules.

Hydrophobic groups are insoluble in water but soluble in nonpolar solvents.

Hydrotropy the process by which the water insoluble substances are made into water-soluble substances.

Laser the emission of increased light intensity in accordance with incoming waves by the excited atoms.

Mass spectrum the spectrum produced by the separation of molecules based on their masses.

Membrane potential the potential difference across the membrane due to the differential distribution of anions and cations.

Monochromatic light the electromagnetic radiation with a single wavelength.

Monochromators dispersion elements that disperse radiation based on wavelength.

Noncovalent bond a weak attractive force, but which determines the properties and functions of biomolecules.

Nonpolar molecules   neutral molecules with electrical symmetry where the positive and negative centres coincide without having dipole moment.

Nuclear resonance   the change of electromagnetic spectrum from low-energy state to high-energy state in the radiation region by the nucleus in a magnetic field.

Nucleosomes   bead-like structures formed by the combination of DNA and histones.

Optical rotatory dispersion (ORD)   the differential rotation of light, depending on different wavelengths.

Osmometer   an instrument used to measure the osmotic pressure of dilute solutions.

Osmosis   the movement of solvent molecules from a region of their higher concentration to a region of their lower concentration through a semipermeable membrane.

Osmotic pressure   the hydrostatic pressure required to prevent the flow of solvent molecules through a semipermeable membrane.

Phase change   the retardation of light waves when the light ray is passed through an object that has a high refractive index.

Photocell   a device that produces electric current in proportion to the intensity of radiation and so useful to detect and measure the light.

Photomultiplier tube   an instrument that consists of a number of successive dynodes so that a large number of electrons are produced in proportion to the falling radiation.

Photovoltaic cell   one that produces electrical voltage in proportion to the intensity of the incident radiation.

Polar molecules   molecules that possess electrical polarity due to positive and negative charges separated by a small distance.

Polymerization   formation of repeats by the combination of two or more monomers resulting in a macromolecule.

Protein sequencing   refers to the determination of the primary structure of a protein molecule.

Protein targetting   the transportation of proteins to different regions of a cell.

Radioactivity   the process of emission of electromagnetic radiation by the radioactive isotopes.

Raman effect   the possession of additional frequencies by the scattered light than that of the incident light rays.

Rf value   the ratio of distance travelled by the sample to the distance travelled by the solvent front.

Surface tension   the force in dynes acting at right angles to any imaginary line of 1 cm length on the surface.

Viscosity   the resistance of a liquid or gas to flow due to frictional effect of one layer on another and is one of the transport properties of liquids and gases.

REFERENCES

1. *Cell and Molecular Biology*, Ed. P. De Robertis and E. M. F. De Robertis. 8th Ed., B. I. Waverly Pvt. Ltd., New Delhi, 1995.

2. *Text Book of Physical Chemistry*, S. Glasstone. 2nd Ed., Macmillan & Co. Ltd., London, 1956.

3. *Handbook of Isotopic Tracer Methods*, Sakami, W. Western Reserve School of Medicine, USA, 1965.

4. *Manometric and Biochemical Techniques*, W. W. Umbriet, Z. H. Burri and Stauffer, J. F. 5th Ed., Burges Pub. Co., Minneapolis, 1972.

5. *Biophysical Chemistry: Principles and Techniques*, A. Upadhyay, K. Upathyay and N. Nath. 3rd Ed., Himalaya Publishing House, 2003.

6. *Molecular Cell Biology*, Lodish et al. W. H. Freiman and Company, New York, 2004.

7. *Biophysics: An Introduction*, R. M. J. Cottenill. John Wiley & Sons Ltd., England, 2002.

8. *Medical Biophysics*, R. N. Roy. Books and Allied (P) Ltd., Calcutta, 2001.

9. *An Introduction to Physical Biochemistry*, H. B. Bull. 2nd Ed., F. H. Davis, Philadelphia, 1971.

10. *Biophysical Chemistry*, Cantror, C. R. and P. R. Samuel. Framan, 1980.

11. *Electrodes and the Measurement of Bioelectric Events*, Geddes, L. A. John Wiley & Sons Inc., New York, 1972.

12. *Principles of Protein Structure*, Schultz, G. E. and Schirmer, R. H. Springer–Verlag, 1973.

13. *Physical Chemistry of Macromolecules,* Tanford, C. Wiley, New York, 1961.

14. *Structure and Stability of Biological Macromolecules,* Timasheff, S. N. and Fasman, G. D. (Eds.). Dekkar, New York, 1969.

15. *Physical Biochemistry,* Van Holde, K. F. Prentice Hall, N.J., 1971.

INDEX

A

action potential 281
    definition 282
    monophasic 289
    diphasic 290
adsorption 103, 106, 122, 127, 198,
    201, 207, 217, 222
adsorption isotherm 105
antibodies 70, 72
antigens 70, 71, 72, 144, 227, 228
autoradiography 4, 68, 297, 301,
    303, 305

B

Beer–Lambert's law 152
bioelectricity 279, 291
biomolecules 3
buffer solution 95
buffer systems 183
    bicarbonate 184
    phosphate 184
    protein 185
    amino acid 185

C

cathode ray oscilloscope 287,
    291, 292
centrifuges 191
    desktop centrifuge 192
    high speed centrifuge 192
    ultracentrifuge 192
centrifugation 189
    basic principle 189
    relative centrifugal force 190
    preparative 193
    differential 193
    density gradient 194

rate zonal 195
isopycnic 196
analytical 196
chaperones 33
chromatography 198
    principle 198
    interactions 199
    locating reagents 204
    solvent systems 202
    paper 200
    two-dimensional 205
    adsorption 200
    thin layer 200
    column 207
    gel permeation 209
    ion exchange 210
    affinity 211
    GLC 212
    HPLC 213
circular dichroism 267, 269, 275
colloids
    Tyndal effect 120
    Brownian movement 120
    emulsion 121
colorimeter 157
covalent bonds 12–16, 32, 34, 42,
    44, 47, 49, 53, 55, 74, 103, 145,
    172
crystal counter 305

D

DNA
    primary structure 61
    secondary structure 61
    sequencing 65
denaturing agents 44, 75
depolarization 280, 282, 291
dialysis 94–98, 119, 127

diffusion
    definition 79
    Fick's law 82
    potential 83
    facilitated 83
dipole 7, 42, 74
dipole interactions 9, 10, 74

E

effective filtration pressure 94
electromagnetic radiation 233, 234
electromagnetic radiation 4, 145, 233–237, 242, 253, 258, 262, 295
electromagnetic spectrum 234, 236, 258, 262
electrophoresis 215
    principle 215
    supporting media 217
    boundary 218
    paper 219, 220
    PAGE 222
    cellulose acetate 222
    SDS – PAGE 225
    slab gel 225
    immunoelectrophoresis 227

F

filtration 92–94, 127, 200, 209
filtration fraction 94
flame photometer 160–163
furanose form 46

G

Gibb's adsorption equation 105
glycoproteins 24, 51, 52, 75, 222
GM tube 297–299

H

H-DNA 65
helmholtz–gouy electrical double layer 107, 128
hydrogen bonds 10, 16, 28–32, 41–45, 48, 62, 64, 74, 109, 199, 225
hydrophobic groups 10, 11, 45
hydrotropy 110, 128

I

integral proteins 23
ionic bonds 15, 30, 42, 44, 45, 74

L

laser 272
    principle 272
    the ruby laser 273
    helium–neon laser 274
ligand interactions 35, 75, 202
liquid scintillation counter 305

M

macromolecules 3, 4
membrane permeability 21, 22, 74, 123
membrane potential 83, 84, 279, 280, 286, 288, 291, 292
microscopy 133
    compound microscope 134
    phase contrast microscope 136
    interference microscope 138
    polarization microscope 140
    ultraviolet microscope 141
    fluorescent microscope 143
    ultramicroscope 144
    electron microscope 145

TEM 146
SEM 148
electron cryomicroscope 149
scanning tunneling electron
microscope 149
myelinated nerve 281, 283, 292

N

nerve impulse 281–285, 291, 292
neuromuscular junction 284,
285, 292
non-covalent bonds 12, 15, 13,
25, 32, 34, 53, 55, 74
non-myelinated nerve 281, 283,
292
nucleoproteins 69
nucleoproteins 75, 142, 143, 302
nucleosomes 69, 70, 75

O

ORD 267, 268, 269, 270, 271, 275
osmometer 88
definition 88
mercury osmometer 88
cryoscopic osmometer 89
osmotic pressure
definition 91
determination 91
osmosis 21, 85, 86, 87, 88, 127

P

pH 169
kinetic theory of pH
pH scale 170
pH indicators 173
Henderson–Hasselbalch
equation 173

pH meter 175
principle 175
reference electrodes 177
glass electrodes 179
phospholipids 18, 19, 55, 56, 122,
268
photometry 151, 153, 154, 160
polymerization 3, 12, 68, 74, 117,
223
precipitation 111, 112, 122, 128
protein 26
primary structure 27
secondary structure 28
tertiary structure 30
quaternary structure 32
folding 32
interactions 34
sequencing 38
denaturation 41
pyranose form 46, 47

R

radioactivity 4, 295–297, 300, 301,
304
radioisotopes 4, 68, 295, 296
repolarization 283
repolarization 280, 282, 283, 289,
291

S

saltatory conduction 283
scintillation counter 299, 305
spectrophotometer 158, 159, 162
spectroscopy 235
basic principles 236
ultraviolet and visible 237
infrared 239, 240
fluorescent 242

atomic absorption   243
atomic emission   245
mass   246
raman   249
X-ray absorption   252
X-ray diffraction   253
NMR   258
ESR   260
stability of macromolecules   12
surface tension   98–106, 111, 127

synapse   284

V

van der Waal's interactions   14, 17, 18, 19, 56, 74
Vant Hoff's theory   86
viscosity   11, 17, 44, 113–117, 128

Z

Z-DNA   64